D0849678

DYNAMOS AND VIRGINS REVISITED:
WOMEN AND TECHNOLOGICAL CHANGE IN HISTORY

An anthology

edited by

Martha Moore Trescott

THE SCARECROW PRESS, INC.
METUCHEN, N.J., & LONDON
1979

Library of Congress Cataloging in Publication Data
Main entry under title:

Dynamos and virgins revisited.

    Includes index.
    1. Women--History--Addresses, essays, lectures.
2. Women--Social conditions--Addresses, essays,
lectures. 3. Technology--History--Addresses, essays,
lectures. 4. Women--Employment--History--Addresses,
essays, lectures. I. Trescott, Martha Moore, 1941-
HQ1154.D95   301.41'2'09   79-21404
ISBN 0-8108-1263-0

Copyright © 1979 by Martha Moore Trescott

Manufactured in the United States of America

CONTENTS

INTRODUCTION

The topic of women in technological history has become an important area of research, writing and teaching in the last few years. In 1976 the organization, Women in Technological History (WITH), a sub-group of the Society for the History of Technology, was founded. In less than one year, WITH nearly quadrupled its mailing list to those with active interests in this area. This rapidly growing subject field, focusing on both current and historical developments, is generating its own bibliographies and the need for review essays. [1]

Because of increasing interest and heightened awareness of the need for both synthesis and analysis concerning women in technological history, this anthology has been prepared. While it can only serve to introduce the field, it is hoped that the book will stimulate much further research and analysis. It was thought especially desirable at this time to collect some of the more significant, pioneering works in this field, as they have been quite scattered throughout the journal and symposium literature.

Dynamos and Virgins Revisited was chosen as the title of this book for several reasons. First of all, the words symbolized women and technology in history in that the image of a particular woman in people's minds (the Virgin Mary) helped fuel the construction of cathedrals in Europe several hundred years ago. Second, the editor's own essay on "Julia Hall and Aluminum" literally deals with the contributions of a spinster to electrochemical technology involving dynamos! Finally, Lynn White, Jr. utilized the title derived from Henry Adams in an earlier anthology on technological change, Dynamo and Virgin Reconsidered, Essays in the Dynamism of Western Culture (originally published as Machina ex Deo), which contained one essay very relevant in many ways to women's history as well as to the history of technology: "The Necessity of Witches." For these reasons, as well as for our belief that commentaries on our

1

technological society, past and present (such as those by Adams and White), are in need of revisions which take into account approximately one-half of the population, this title was selected.

Four of the essays in the anthology derive from oral presentations before the Society for the History of Technology (SHOT) in Philadelphia in December, 1976. These were brought together for SHOT by this editor in a session entitled "Women and Technological Change: Some European and American Perspectives." (Four of the other papers in this collection were also presented, in some form, before SHOT, two in 1973 and two in 1975.) The title of the 1976 session captures the breadth of scope of the session, for the papers dealt with women as participants in and contributors to technological change as well as those upon whom technology was impinging--and sometimes with both at once. Also, the papers were not limited to a particular time period or geographical area. Thus developments in both Europe and America were considered relevant, although both that session and this anthology deal primarily with American history. Topics discussed ranged over approximately two hundred years, mainly focusing on events of the eighteenth, nineteenth and twentieth centuries. The session's chairperson, Ruth Schwartz Cowan (author of two essays in this volume) wisely did not attempt to impose any one major integrating theme on the papers at that time, and I will follow her lead in this anthology. However, Professor Cowan did suggest several important themes and ideas around which women's history of technology might cohere, and I will build upon certain of those, adding Gestalten of my own.

The papers collected here deal with several main areas of the historical interaction of women and technology. First of all, Part I(A) deals with women as active participants in technological change. The treatment of the topic of women as operatives and workers in industry ranges from the eighteenth-century French silk industry (Daryl M. Hafter) to nineteenth-century American manufacture of carpets (Susan Levine) and paper (Judith A. McGaw). Women as inventors, engineers, scientists and entrepreneurs are considered in Part I(B) in papers by Deborah Jean Warner, Margaret W. Rossiter, and Martha Moore Trescott.

The second major part of the anthology deals with some effects of technology on women in the more private spheres of life: in the home and in reproduction, child-raising and socialization. On technology and women's work in the home, Susan J. Kleinberg and Ruth Schwartz Cowan examine aspects, respectively, of certain working-class women before the twentieth century and of middle-class women thereafter. Both are concerned mainly with U.S. conditions. Part II(B), also treating aspects of women's life in the home, includes essays by Vern L. Bullough and Carroll W. Pursell, Jr., who consider certain ways in which technology has affected women as child bearers and raisers, with larger sociological implications. Bullough considers the technologies of contraception, nursing, and the sanitary pad; Pursell views sexism as historically propagated by toys.

The arrangement outlined above follows somewhat the analysis set forth in Ruth Cowan's paper, "From Virginia Dare to Virginia Slims: Women and Technology in American Life." It was thought important to place this paper first in the anthology, primarily because it establishes various themes for viewing women and technological change. It therefore stands alone, not as an element of either main part, serving rather as an introductory piece to help guide the reader through the subsequent essays.

In "From Virginia Dare to Virginia Slims," Professor Cowan has articulated four major topical areas which need to concern women's historians of technology: "women as bearers and rearers of children," "women as workers," "women as homemakers," and "women as anti-technocrats." She contends that women's history of technology is entirely justified as a separate field from men's history of technology, for several reasons. One is that "women menstruate, parturate, and lactate; men do not. Therefore, any technology which impinges on those processes will affect women more than it will affect men," as explored in part in her essay in Part 2(A). [2] In addition, in considering women as workers in the market economy, "women workers are different from men workers," because (1) "when doing the same work women are almost always paid less than men; (2) considered in the aggregate women rarely do the same work as men (jobs are sex-typed); and (3) women almost always consider themselves, and are considered by others, to be transient participants in the workforce." [3] Further, "both men and women live in homes, but only women have their 'place' there." [4] And finally, women have been conditioned historically to feel that they cannot comprehend technology, that indeed technical matters constitute a male realm. [5] (Pursell's essay is partly relevant here.)

The papers in this anthology generally support Professor Cowan's analysis. In the first part of the book, the papers by Hafter, Levine and McGaw indeed indicate that women as workers and industrial operatives, whether in France or America, were in part employed because they were a source of cheap labor. Further, they were often subject to sexist abuses spurred by technological change. Hafter has discovered much anti-female propaganda to justify displacement of the drawgirl by the Jacquard loom, while Levine notes sexist propaganda directed at the female weavers who displaced males as the power loom became adopted in American carpet-making. It is interesting to consider Professor Cowan's remark that work has been sex-typed, in the light of technological changes depicted in these essays. Levine's paper indicates that with the advent of mechanized weaving, less-skilled female workers displaced the more-skilled male weavers, and Cowan notes that in cigar-making in the final decades of the nineteenth century the replacement of skilled males by less-skilled females accompanied the introduction of machines. Hence, in these instances, although females displaced male workers, the character of the task became quite different from the work earlier performed by men. Indeed, in general the women workers have historically been considered the less skilled, as Judith McGaw shows in her essay, and the more menial, repetitive and monotonous work has fallen to them, whether machine work or by hand. In fact, McGaw's paper shows that women's work in the nineteenth-century Massachusetts paper industry was directly affected very little by mechanization, since the machines were generally operated by men. Also, as Cowan notes, women have historically been paid less than men, in general, whether as skilled or unskilled workers (and one should include here white-collar and professional labor, too). [6]

Professor Francine Blau has elsewhere presented data on "Women in the Labor Force: The Situation at Present," viewing trends up through 1974. She has found that in general women made only 57 per cent of what men earned in 1973 and that occupational segregation by sex is even more severe and entrenched than racial segregation. According to her index of segregation, there has been little change since 1900 in occupational segregation for women; this supports Cowan's statement that women and men rarely perform the same types of work. [7] Prior to Blau's study, Robert Smuts had investigated women in the labor force, focusing primarily on the data through the 1950s (with valuable historical commentary). He found that "in broad outline ... the picture of women's occupations outside the home has changed since 1890 in only a few essentials,"[8] and that

> In recent years employers have been able to attract millions of additional women into the labor force without changing the relative levels of men's and women's pay, or greatly expanding women's opportunities for advancement. The decade between 1945 and 1955 was one of booming prosperity, labor shortages, unprecedented peacetime demand for women workers, and unprecedented increase in the number of women working. Yet, in 1955, the median wage and salary income of women who worked full time was still less than two-thirds that of men--almost exactly the same as it was in 1945.[9]

In the mid-1970s, according to Blau's figures, the gap between men's and women's pay was even greater than twenty or thirty years earlier and, further, the gap is ever widening.[10]

Earlier, in the 1920s, according to Irving Bernstein, "women entered the labor force at an accelerated pace" and by 1930 "almost one of every four persons in the labor force was a woman."[11] Female high school graduates in increasing numbers expected to work for pay after graduation. Yet

> the growing number of women entering the labor force could count upon unequal treatment in wage rates. The NICB [National Industrial Conference Board] survey of manufacturing, for example, found that the average hourly earnings in 1929 of male skilled and semiskilled workers were 67¢, of male unskilled 50.3¢, and of females at all levels of skill 40.1¢. In foundries in 1929 the earnings of male coremakers were 74.4¢ and female 46.9¢, male laborers 49¢ and female 38.6¢, male assemblers 65.7¢ and female 44.1¢. In meat packing in 1929, earnings of male trimmers were 52.1¢ and female 37.1¢, male blowers, graders, and inspectors 51.7¢ and female 38.4¢, male machine tenders 53.1¢ and female 35.4¢. In the furniture industry in 1929, male assemblers earned 56¢ and female 31.7¢, male craters and packers 43.5¢ and female 33.1¢, male finishers 50.5¢ and female 37.1¢. In cotton goods, earnings of male frame spinners in

> 1928 were 33.9¢ and female 27.6¢, male creelers
> 29.8¢ and female 23.9¢, male weavers 39.2¢ and
> female 37.1¢.
> ... [T]he spread widened during the twenties,
> if the NICB survey is a guide. In 1923 the dif-
> ferential in manufacturing between the average
> hourly earnings of the male skilled and semiskilled
> over females of all grades of skill was 22.8¢; by
> 1929 the gap had broadened to 26.9¢. In 1923 the
> male unskilled enjoyed an advantage of 6.3¢ over
> females of all skills; in 1929 the spread reached
> 10.2¢. The rule in American industry was unequal
> pay for equal work based upon sex. [12]

In addition, the many studies on women and work
which have been compiled into a bibliography by Martha Jane
Soltow and Mary K. Wery, American Women and the Labor
Movement, 1825-1974, and also the book America's Working
Women, A Documentary History--1600 to the Present give
ample evidence concerning patterns of discrimination against
women in the labor force. One important conclusion which
might emerge from further study of some of these works,
especially those in Soltow and Wery dealing with mechanization of industry, is demonstrated in Susan Levine's essay in
this anthology: that is, as some industries such as textiles
and cigar-making have become more mechanized, women
workers have tended to displace male workers. In this situation, the men would often vilify the women rather than the
machine. [13] Also, it is interesting to note that regardless
of the extensive technological changes in the last 200 years,
women have not achieved "equal pay for equal work."

Concerning the second major area of women as "prime
movers" of technological change in this collection (Part I B)--
women as inventors, engineers, scientists and entrepreneurs
--the reader will see women not as "anti-technocrats" (as in
Professor Cowan's terminology) so much as persons who contributed to the on-going technological system. However, the
papers by both Rossiter and Trescott certainly uphold Cowan's
ideas about discrimination against women in scientific and
technical fields. Much more taxonomic and biographical work
needs to be done before we can know the extent to which
women as scientists and engineers have participated in technological change, even as Rossiter indicates. [14] However,
women in other, more traditional roles, such as performed
by Julia Hall (see Trescott essay on Hall) and undoubtedly by
the scientist wives of, for example, the zoologists noted by

Rossiter, certainly must have contributed a great amount of energy and skills to their male relatives' endeavors, especially when the separation of invention and business, on the one hand, and home were not as severe as today.

However, it is interesting to note that, even with the rise of the factory system, which tended increasingly to separate some jobs from the home, much technological change and industrial and economic growth were literally built upon the backs of women. Professor Paul Uselding, for example, has mentioned that because the differential between men's and women's wages in early nineteenth-century U.S. textile factories was greater than that for the British analog, growth in U.S. textiles output could compete favorably with and even outstrip British production in a relatively short time by means of employment of women.[15] Thus, as Edith Abbott has also commented, to women operatives in industry can be attributed much of the industrial superiority of the U.S., both in the nineteenth and twentieth centuries.[16]

That women have participated in technological change and industrial growth in many ways is undisputed. Yet the ultimate decision-makers "at the top" have historically been, and still are, men. Margaret Hennig and Anne Jardim, in The Managerial Woman, show that today although women comprise approximately 40 per cent of the labor force, fewer than five per cent earn $10,000 or more per year and only slightly over two per cent of working women make annual incomes of $25,000 or over, figures quite consistent with Blau's findings.[17] It should be stressed, therefore, that in Western society women have not been the real technocrats, as Professor Cowan notes. (It may well have been different in the Egyptian, Celtic, and other ancient civilizations, but much more work needs to be done to determine this.)[18] It could be conjectured that our culture has never before experienced, prior to today's growing "alternative" or "appropriate technologies" movement, technological change directed and determined to a major extent by women. Experiments in construction of a life system based on smaller scale and more ecologically sound technology, such as the New Alchemy, are to be commended.[19] Although the results are by no means known yet, it may well be, as Cowan says, that "as more and more women begin to play active and powerful roles in political life, we may be surprised to discover the behavioral concomitants of the unspoken hostility to science and technology that they are carrying with them into the political arena."[20]

The so-called "negative" view of women as participants in technological change, and the "positive" school
which points to women in active roles as having helped to
produce today's technological order can be reconciled mainly
in two ways. First, both would agree that, in general,
women have not historically been in key decision-making
roles about the course of more modern technology, and that
those women who did participate, whether as operatives, inventors, or scientists and engineers, were discriminated
against in various ways on the basis of sex.

The second major part of the anthology deals with the
interaction of women and technology in the more private endeavors, mostly in the home as wives and mothers. This
placement of essays on the work of housewives and homemakers is not meant to exclude housework from the domain
of "women and work," discussed in Part I. It was felt important, however, to differentiate between women whose work
was either paid labor or connected somehow to market production, on the one hand, and women whose work was restricted to maintenance of the home and family, on the other.
Also, no implication is intended that the homemakers did not
actively participate in technological change and economic
growth, for both as consumers and producers they certainly
did. Simply because the Gross National Product has historically not included the services of housewives explicitly does
not mean that they were an inconsequential part of the growth
of the economy. In fact, in 1921, when national income ac-
counting as a technique was beginning to take shape, it was
estimated that the services of housewives probably comprised
approximately one-third of the total national product, an esti-
mate based only upon the wages of domestic workers--whose
hours were much more limited than those of homemakers. [21]
As a chapter in women's history, it would be extremely
interesting and important to investigate the history of the
compilation of the GNP, asking which men were in charge
of estimating figures for the service sectors and why such
estimates as those given in 1921 were ignored in the GNP
which has ultimately become established. [22]

When the separation of job and home is discussed, we
must bear in mind Professor Cowan's observation that while
both men and women live in homes--even if they hold outside
jobs--housework has historically been and continues to be
primarily left up to women. And Cowan's assertion that the
number of hours per week a full-time, middle-class housewife may spend in domestic chores has increased since 1900

tends to indicate that technological change in household technology may not have "liberated" women as much as is sometimes claimed. Indeed, a major way this has worked has been through technological displacement of domestic servants such as laundresses, cooks and cleaning ladies, resulting in the concentration of housework in the hands of one person, the housewife.

Formerly, middle- and upper-class women served more as managers of an array of household servants, rather than as performers of each task themselves. Elizabeth Janeway has noted this phenomenon in medieval and later times,[23] and it certainly persisted in many areas into the late nineteenth and early twentieth centuries. But as the present century has worn on, the housewife, especially in the U.S., has often become a "Jill-of-all-trades" with a proliferation of technology to support her multi-faceted life.

Also during this century, increasing numbers of women have entered the paid labor force. As technological change in housework has displaced domestic workers, technological change in the business and industrial world has tended to augment demand for female workers such as secretaries and telephone and key punch operators. In fact, it is interesting to juxtapose employment data for female domestic servants and female clerical help and, with the aid of regression analysis, to try to determine the degree of correlation in the decline of the former employment with the rise in the latter. The Occupational Progress of Women, 1910-1930 by the U.S. Women's Bureau in 1933 suggests that perhaps a kind of substitution effect may have operated here.[24] In his first chapter, Robert Smuts also notes the decline in the number of domestic servants and the rise in the number of female office workers, evident even by 1890.[25] Table I depicts this trade-off effect for female workers.

A simple regression between %F for domestic servants (S) and %TF for clerical workers (C) (as noted in Table 1) is given in equation (1), $C = 38.5 - 1.157S$, whose R^2 is .951. However, a simple time trend for C (2) yields a slightly better R^2 of .955. So if both S and T are incorporated, the relationship (3) $C = .203T - .597S + 22.7$ gives an R^2 equal to .974.[26] Therefore, the data are consistent with the view that these two trends are related to each other.

On the face of it, it would not seem plausible to argue

TABLE 1

Employment of Domestic Servants and Clerical Workers in the U.S., 1900-1970
(000's persons)

	Domestic Servants				Clerical Workers			
	Male	%Total M Workers	Female	%TF	Male	%TM	Female	%TF
1900	53	0.22	1,526	28.7	665	2.8	212	4.0
1910	67	0.22	1,784	24.0	1,300	4.4	688	9.2
1920	51	0.15	1,360	15.7	1,771	5.3	1,614	18.7
1930	89	0.23	1,909	17.8	2,090	5.5	2,246	20.9
1940	135	0.34	2,277	18.1	2,282	5.8	2,700	21.5
1950	78	0.18	1,414	8.6	2,723	6.4	4,408	26.7
1960	65	0.14	1,752	7.9	3,024	6.6	6,407	28.7
1970	38	0.08	1,166	3.8	3,748	7.6	10,461	34.5

Source: Historical Statistics of the U.S., Colonial Times to 1970, Part 1, pp. 139-40.

that those females who were or would have been domestic servants increasingly opted for clerical work, because the character of and preparation for the tasks seems different. The clerical labor force would have to be at least literate and, therefore, somewhat more educated than domestic servants in general. However, at least two important considerations enter into the analysis of this trade-off. One is that during the nineteenth century and into the twentieth, more and more women were receiving an education which would have enabled them to function in an office and at a typewriter. Because of a lack of minimal schooling necessary for such tasks, their earlier counterparts may have had less choice about their mode of employment. Second, W. Elliott Brownlee has reminded us that historically there has tended to be little difference in the wages paid to literate and illiterate women. [27] Thus those females who had by 1880 or so become more literate and well-educated than their older counterparts in the labor force might well have had few aspirations of the Horatio Alger type and might have seen their employment choices as restricted to dead-end, low-paying jobs such as clerical work and domestic service. For the literate, surely office work was more prestigious than domestic service.

However the case may be, surely one of the great transformations in the world of business has been the almost total conversion of clerical work from a male-dominated to a female job in the late nineteenth century and the early decades of the twentieth. Although several recent works have dealt with the roots of the close historical association between women and the typewriter, much more research is warranted here. [28] There is also work in progress on the influence of the domestic service industry on the rise of other service sectors in the U.S. in the nineteenth century.[29] (Of course, much clerical work was service work.) With the coming of typewriter technology, employment opportunities for women widened, even though the office job for women was not (as it had been for men) a path to management.

Both today and historically, the U.S. Census categories of manual workers, service workers, and clerical workers typically represent the occupations which contain the largest numbers of female workers. Noting changes over the period 1900-1970 in the percentage of the total male and female work forces registered as professional, technical, and kindred workers, one sees the percentage of male workers here rises from 3.4 in 1900 to 14.0 in 1970, while the percentage of females in this category rises from 8.2 in 1900 to 15.3 in 1970. While at first it may seem surprising that women were as well represented in this category as they have been, it is less so when one realizes that teachers are included here. (And, of course, elementary and secondary school teaching has become an overwhelmingly female job.) The hard core of occupational discrimination on the basis of sex really surfaces when changes in the managerial category during the same period are viewed. In 1900, 6.8 per cent of the total male labor force was classed as managers, proprietors and officials, rising to 10.9 per cent in 1970, whereas women managers comprised only 1.4 per cent of the total female labor force in 1900 and a mere 3.5 per cent in 1970.[30]

At any rate, it is clear that whether in the home or outside, technological change has not enabled women to rise above proletarian status, by and large. It may well be that because technological change in the factories and business sector helped create more jobs for women (who, it should be remembered, were considered cheap, legal labor when competitive child labor was outlawed in this century), more technology for housework was demanded to help potentially shorten the time it takes working women to accomplish household tasks. Yet society continued to uphold the notion that housework was still the women's domain, regardless of whether women worked in outside jobs. There was little real shift in the attitudes of men toward assuming any real responsibility for housework, even though their wives were working in paid jobs in ever greater numbers. Thus technological innovation was more generally substituted for the less ubiquitous full-time housewife than was men's domestic labor. In all of those ads for better detergents and vacuum cleaners, one would seldom see men pushing mops and running sweepers!

So, as Cowan says, housework with the new technologies may well consume for the full-time housewife more time than for her turn-of-the-century counterpart. And for the working woman, such technologies can help in certain tasks in the absence of societal values which would encourage men to move into the household sector. In either case, such technological change in the home was geared to the services of women, not men, and tended to keep women in their "place."

With the rise of modern suburbia, with the increasing separation of job and home made practicable by the rise of transport technology such as, first, the electric railway and bicycle and, later, their counterparts, the commuter train and automobile, woman's "place" became more and more a ghetto. Modern feminist writers such as Phyllis Chesler and Betty Friedan have noted some of the stressful psychological effects of woman's ghettoization in suburban homes, surrounded only by children, schools and shopping centers.[31] As Scott Donaldson has written, suburbia has been greatly condemned by sociologists and others, especially since World War II (although Donaldson feels these attacks are largely unjustified, if one views the reality of suburbia and not the myth which it is supposed to live up to).[32]

One thing which may be said with certainty about the suburban home is that it is not the hub of the kind of work activity which was historically carried on earlier in the homes of the well-to-do and others in Europe and America, in both rural and urban areas. As noted before, the home was once the center of work carried on by domestic servants, as well as the wife and children, and might well have been the place of business of the husband (not only farmers and ranchers with their often numerous hands but also artisans with apprentices and those with small business enterprises). The suburban home of the twentieth-century U.S., however, is still the woman's place but is no longer respected as a place of legitimate work. Indeed, as Ruth Cowan has said in her essay, published elsewhere, "Two Washes in the Morning and a Bridge Party at Night: The American Housewife Between the Wars":

> In earlier years American women had been urged to treat housework as a science; now they were being urged to treat it as a craft, a creative endeavor. The ideal kitchen of the prewar period had been white and metallic--imitating the laboratory. The ideal kitchen of the postwar period was color coordinated--imitating the artist's studio. Each meal prepared in that ideal kitchen was a color composition in and of itself: 'Make Meals More Appetizing by Serving Foods that Have Pleasing Contrast of Color.' Ready made clothes could be disguised by adding individual hand sewn touches; patterned towels could be chosen to match the decorative scheme in the bathroom; old furniture could be repaired and restyled. The new housewife would be an artist, not a drudge.[33]

This propaganda, which accompanied technological change in the household sector, tended to change the image of the home substantially--from a place of respected work to one of fun, leisure and even recreation. The image of the housewife suffered. Cowan has noted the effects of some advertising, 1920-1940, Friedan has criticized later advertisers, and both authors have focused on the manufacture of the "feminine mystique" of this century. Indeed, the feminine mystique has at its core an image of woman as imbecile, "happily" pushing mop and vacuum cleaner.

Of course, popular compendia of the nineteenth century advised "happy" housewives to "sing merrily while you build the nest."[34] Indeed, Charles M. Hall himself (see Trescott essay in Part I B), as a youthful salesman of the Golden Censer, was set fairly straight by one such "happy" homemaker he called on in 1882.[35] It may be that the mystique as such could not make substantial popular headway until the rise of modern communications systems such as radio and television, and any women's history of technology should certainly study the effects of these monumental technological innovations on the image of women. Radio and especially television have certainly combined, through both advertising and programming, to effect an imbecilic image of women, particularly as homemakers, significantly degraded from the matron of the middle-class home who earlier was a manager of various servants and who may well have had more contact with her husband's work during a given day.

Carroll Pursell has also explored aspects of the period 1920-1940 and shows that toys began to reflect the "technical marvels" of the age, including the new household technology. And through this medium, little girls early in life became socialized to the new image of housewife. Female children could have "Raggedy Ann ironing toys and a play kitchen. Each of these Queen Size items is so sturdily constructed, we are told, that it 'makes a little girl want another Queen Size appliance.'"[36] The linkages between adult technology and toys, between both adult and children's technology and advertising, between advertising and media technology, effected a monolith which socialized females away from careers in science and engineering. It is intriguing to consider that the various kinds of media executives, writers, editors, engineers and technicians, including camera personnel and photographers, have typically been men, as the various mass images of twentieth-century women--from inane housewife to pornographic model--have become so entrenched. It would not be unfair to say that both media personnel and the technology they have used have combined to degrade the image of women in many ways. W. Elliott and Mary Brownlee have noted the seeming paradox between women's rising participation in the labor force since World War II and "an intensifying cultural fixation on their virtues as housekeepers, child rearers, and husband custodians," and that after 1910 the "immediate effect of the technological revolution in the household was not to increase the size of the female labor force but, rather, to encourage even greater specialization of women in the tasks of child nurturing."[37] The media have undoubtedly played a large role here.

Along with the media, the automobile has contributed to the isolation of women and children in the suburban home, to the limited contact of a homemaker with "working" adults during the day, and to mental depression in women whose major work centers in the home. The automobile has further helped increase the number of hours per week a wife and mother can spend in home-related chores, since she has become a family chauffeur. Women have always been a major market for the automobile, even as they were perhaps one of the most crucial markets which sustained the bicycle boom of the nineties (which, incidentally, initiated many of them as drivers of self-propelled vehicles that were important precursors of automobiles).[38] Yet the automobile, like other technologies inherent in the media, household, industrial, and business innovations already discussed, has often kept women in subservient roles. Although the automobile has increased the mobility of both women and men and has facilitated transport to paying jobs outside the home, social values and customs allowing males more generalized mobility than females, in the labor force and in society at large, have changed very slowly.

It is also interesting, as Vern Bullough's paper shows, that Western society has for decades had adequate technologies to aid contraception, nursing and menstruation but that there has characteristically been quite a cultural lag in terms of change in societal attitudes toward men as child-raisers. A similar lag has been noted in the area of men as homemakers in general. And in his discussion of sexism in toys, Pursell has demonstrated, once again, that technology has been made to serve and propagate social attitudes.

In summary, it would seem that women's history of

technology might well shed some light on one of the largest
of the "big questions" in the history of technology, in George H.
Daniels' phrase. [39] Specifically, does technology determine
the directions in which society moves, or does society more
nearly determine what kind of technological change it wants,
how to bring that change about, who will administer such
change and what purposes the new technologies will serve?
In other words, is technology a mindless monolith which dic-
tates to us or do social values determine the path of technological "progress"? Melvin Kranzberg has claimed that
"technology is neither good nor bad; nor is it neutral," and
has focused on the decision-makers who help determine technological change. [40] David Landes in the Unbound Prometheus
has depicted a kind of technological momentum, and Eugene
Ferguson has cited the technological imperative. [41] Recently,
a student of Ferguson's, Thomas C. Guider, has provocatively
questioned technological and economic determinism in the rise
of electrical interconnection earlier in this century, asking
who in the "establishment" supported such interconnection. [42]

As an increasing number of consumer advocates and
others are asking, who is in charge of decision-making about
technological change and how does the "average" person help
register her/his feelings about the nature and direction of
technology? Women's history of technology has these very
questions at its core. Who have been the decision-makers
who have told women what they as consumers wanted in the
technologies of the household, reproduction and menstruation?
Why, women are beginning to ask, are most birth control
technologies women's devices, noting the lag in societal attitudes toward responsibility for child-bearing? The fact that
the technological establishment has discriminated against its
female wage-earners and professionals would seem to be on
a continuum with the fact that technological change has tended
to keep women in a subservient status, either in the home or
out. And, of course, women's history of technology, then,
must ultimately question the usage of "progress" as applied
to modern technological change. For it would not appear that
our modern technological society has seen as many improvements as might have been expected in the status of its women.

Improved standards of living, of course, have
benefited many. Yet a growing number of families in
the U.S. are headed by females (one in eight of all
the families in the populace, as of March 1975, a
percentage that is continually increasing), and one half
of these families live at or below the poverty level,

as compared with less than one-tenth of male-headed families.[43] Certainly, even poor families in the U.S. tend to be better off than their counterparts elsewhere, so it could be argued that technological and economic growth in the U.S. have indirectly made life better--but not selectively better for women, and certainly not for the status of women.

It also might be claimed that among the benefits to women of our modern technological order have been the advances in modern medicine and sanitation which, among other things, have helped reduce the frequency of maternal death in childbirth. It should perhaps also be borne in mind that, as Robert Fogel has said, women have generally always lived longer than men, although the mortality of women of child-bearing age has been substantially reduced in this century.[44] However, the oppression of women by the medical establishment, in prescription of harmful birth control measures, such as intra-uterine devices and hormones, must also be weighed in the balance. And, added to that, the virtual exclusion of women as medical practitioners in modern medicine, when once they were considered reliable doctors, should be noted, along with the severe segregation of women in medicine into the nursing profession and as medical and dental technicians. (Why, it should be asked, if women are so adept at dental technology, are there not more female dentists? It would seem that there are fewer female dentists than female doctors, yet the widely used Historical Statistics, Colonial Times to 1970 does not list dentists by sex.)[45]

What have been the effects of technological change on women, overall? Technological change and mechanization, have, in some instances, at least moved women out of the home and thereby widened their sphere of contacts, if not influence. Yet the Industrial Revolution may well also, as Gerda Lerner notes, have stimulated the rise of the "true woman" cult, thereby widening the social gaps between middle- and upper-income women, on the one hand, and working-class and poor women, on the other.[46] The essay by Susan Kleinberg (Part II A) certainly demonstrates a disparity in the technological aids available to working-class women in Pittsburgh, 1870-1900, and those which their more well-to-do counterparts possessed. Industrialization undoubtedly augmented the ranks of the middle classes and provided more leisure. (Perhaps it enabled more men to "buy off" the women, or at least to promise them financial security upon marriage, a hope which often turned out to have been illusory.) And it is significant that as more men entered the middle classes, higher education could be more and more afforded, for both men and women. Yet marriage has often meant in the past that our educated women would be "lost" and would not be enabled to make good use of their education, whereas it has meant just the opposite for the husbands of educated wives (see Rossiter's paper, Part II B). This does not imply, of course, that the education of these women did not benefit the economy greatly; it only means that their contributions have often gone unrecorded and were subsumed under the achievements of their men (see Trescott essay, Part I B). At any rate, more goods and income provided by the industrial revolution have given rise to more leisure and, concomitantly, to the cult of the "true woman," the "feminine mystique" and, as Thorstein Veblen termed it, the "barbarian status of women."[47]

Has modern technology liberated or oppressed women, all told? Helen Campbell said in 1891 that the factory system in America had evolved over the nineteenth century from the potential liberator of women to their oppressor.[48] We need to try to determine if this is true and, if it is, why. To determine what happened to women operatives, including the effects of technological change, between Lowell and the mass-production assembly lines of Texas Instruments and others today is the business of women's historians of technology as well as of social critics and philosophers such as Simone Weil.[49]

Certainly the effects of the industrial revolution and technological change on women are ambiguous and mixed. It would appear at first sight that technology has not tended to liberate women but, rather, that sex roles and economic, political and legal arrangements between the sexes have remained amazingly stable in the face of massive technological developments which might otherwise have tended toward change in or reversal of sex roles. Cowan, in her overview essay, for example, notes the continuing male dominance of the typesetting trades even after the introduction of the linotype, and many other new technologies with push buttons and hoists should have conduced toward less sex-typing jobs formerly defined as male on the basis of physical strength. Yet change here has been slow. Bullough demonstrates a lengthy lag between technological changes in child-rearing, such as the coming of the nursing bottle, and the willingness of society to allow men a substantially greater role in raising children. Similarly, Cowan, the Brownlees and others have stressed social values and folkways which tended to keep the major responsibility for housework in the hands of women despite the many technological changes in housework in this century.

It would seem that the image of women has, in fact, been considerably denigrated in the media in this century, particularly after World War II with the rise of television. One wonders, then, if the technological changes involved in the rise of movies, radio, television and other mass media forms did not tend to cancel the potentially liberating effects of many of the other technological changes noted above. It may well be that the women's movement of the 1960s and 1970s, touched off as it was in part by Friedan's Feminine Mystique, was a reaction against this very dehumanization of the image and work of women in our present century.

In any case, it is important to realize, as noted earlier, that the technologies, such as those involved in the media, cannot in and of themselves be credited or blamed for the slow rate of improvement in the status of women over time. Rather, this book shows that we must look to the various groups of decision-makers who effect technological change and use the new technologies (the entrepreneurs, the corporate leaders, the engineers, government and the consumers, as Otto Mayr has observed).[50]

As was mentioned by various speakers at the Conference on Critical Issues in the History of Technology in Roanoke, Virginia, in 1978, when one views technological change in modern history from the standpoint of the workers and minorities such as the American Indian, the positivist approach is less easy to support.[51] Similarly, if one views women and technological change, one might well ask: what price technological "progress"? Women's history of technology has already begun to raise and to answer some of our most searing and searching questions about technological change and may prove a valuable tool in the formulation of new directions toward a more humane tomorrow.

Notes

1. WITH was founded on December 29, 1976. The minutes of the organizational meeting appears in Technology and Culture, XVIII (July, 1977), 496-7. Also, a

copy of these minutes and of the first WITH newsletter (June, 1977) and subsequent issues are available
from this author, current secretary of WITH. The
first WITH bibliographical newsletter, edited by WITH
Keeper of the Catalogue, Eleanor Maas, was issued
May, 1977, and is also available upon request. A
bibliography on women and technological history has
been prepared for publication in Technology and Culture,
also by Ms. Maas.

2. Ruth Schwartz Cowan, "From Virginia Dare to Virginia
Slims: Women and Technology in American Life, "
first presented to the Society for the History of
Technology, October, 1975, Washington, D.C., and
now appearing in this volume, pp. 30-1.

3. Ibid., pp. 32-33.

4. Ibid., p. 37.

5. Ibid., pp. 40-41. Patricia Cohen of the University of
California at Santa Barbara has researched the socialization of women historically against widespread
participation in mathematical disciplines.

6. Cowan, "From Virginia Dare to Virginia Slims, " p. 33.
Also see, for example, Helen L. Sumner, History
of Women in Industry in the United States (Washington, D.C., 1910), (Vol. IX of Report on Conditions
of Women and Child Wage Earners in the U.S. in 19
vols., U.S. Senate Document 645, 61st Congress, 2d
Session), cited as number 49 in the bibliography by
Martha Jane Soltow and Mary K. Wery, American
Women and the Labor Movement, 1825-1974: An
Annotated Bibliography (Metuchen, N.J., 1976), pp.
16-17. Also, see Edith Abbott, Women in Industry;
A Study in American Economic History (New York,
1913).

7. Francine D. Blau, "Women in the Labor Force: The
Situation at Present, " to appear in Women and Their
Work, ed. by Ann Yates and Shirley Harkness (Palo
Alto, forthcoming), pp. 20-21 and 24-25 of type-
script (March, 1976), provided the author by Pro-
fessor Blau.

8. Robert W. Smuts, Women and Work in America (New
York, 1974), p. 35.

9. Ibid., p. 108.

10. Blau, "Women in the Labor Force," p. 24 of type-
    script.

11. Irving Bernstein, The Lean Years, A History of the
    American Worker, 1920-1933 (Boston, 1960), pp.
    55-56.

12. Ibid., p. 69.

13. See, for example, item number one in Soltow and Wery
    (cited in full in note 6 above) by Edith Abbott, "Em-
    ployment of Women in Industries: Cigar-Making,
    Its History and Present Tendencies," Journal of
    Political Economy, XV (January, 1907), 1-25, and
    also additional listings under "Displacement" in the
    Soltow and Wery subject index. In America's
    Working Women, A Documentary History--1600 to
    the Present, compiled and edited by Rosalyn Baxan-
    dall, Linda Gordon, and Susan Reverby (New York,
    1976), see items under "mechanized production
    (1920-1940)."

14. Margaret W. Rossiter, "Women Scientists in America
    before 1920," American Scientist, 62 (1974), 322,
    now appearing in this volume, Section I(B). Also,
    Notable American Women is now seeking to update
    its biographies on women in the scientific and tech-
    nical areas. Carroll Pursell is currently research-
    ing women engineers in the U.S. around 1900, es-
    pecially within the home efficiency movement, and
    Alva Matthews of the Society of Women Engineers (SWE)
    has given a talk on pioneering women engineers in
    U.S. history (copies of this talk available through
    Martha Trescott upon request and by permission of
    Dr. Matthews). See also "A History of Women
    Engineers in the U.S., 1850-1975," grant proposal
    by Martha Trescott, summer, 1978, to the National
    Science Foundation and the National Endowment for
    the Humanities, and Trescott's progress report on
    this project, SWE Proceedings, 1979.

15. Paul J. Uselding, essay on "Manufacturing" in the
    forthcoming Dictionary of American Economic
    History, p. 24 of typescript provided this author
    by Professor Uselding.

16. Abbott, Women in Industry, p. 317, for example.

17. Margaret Hennig and Anne Jardim, The Managerial
    Woman (New York, 1977).

18. Typically, the Celts are not even mentioned in economic
    histories and histories of technology dealing with
    European developments. Shepard Clough and Richard
    Rapp skip from the classical civilizations of Greece
    and Rome to the Middle Ages, European Economic
    History, The Economic Development of Western
    Civilization, 3rd ed. (New York, 1970), as does the
    Kranzberg and Pursell, ed., Technology in Western
    Civilization, I (New York, 1967), and Richard S.
    Kirby et al. in Engineering in History (New York,
    1956) seems more typical than atypical in treating,
    first, Egyptian and Mesopotamian civilizations, fol-
    lowed by Greek and Roman engineering and then by
    technology of the Middle Ages. Even though William Mc-
    Neill can say that "Celtic tribesmen were the first to
    master the arts of riding; and their expansion from
    southern Germany over most of northwest Europe
    was facilitated by their resort to cavalry tactics
    (pp. 258-9, The Rise of the West), Lynn White in
    Medieval Technology and Social Change (Oxford,
    1962), in discussing the rise of the foot stirrup and
    mounted shock combat, does not explicitly mention
    the Celts except in one footnote on p. 138, referred
    to from a footnote on p. 8. Turning to page 138,
    we see that "it would seem that the Germans got the
    heavy war-horse from a Celtic people...." Marc
    Bloch, in Land and Work in Medieval Europe (New
    York, 1969), mentions the Celts on p. 168. Yet it
    would seem that to economic historians and historians
    of technology, the history of Celtic achievements like
    the long-bow (which, ironically, the Celts did not
    seem to evolve, as they preferred the sword, ac-
    cording to McNeill, p. 259), "belongs more to the
    history of English folklore than to a serious history
    of English technology" (Carlo Cipolla, Guns, Sails
    and Empires, Technological Innovation and the Early
    Phases of European Expansion, 1400-1700 (New York,
    1965, p. 36).
       Has there indeed been a "conspiracy of silence"
    about the contributions of the Celts, as Elizabeth
    Gould Davis asserts in The First Sex (New York,
    1971)? Whether it has been a conspiracy or not,
    there has been silence. Nora Chadwick, perhaps the
    foremost Celtic scholar, has said that "we have been

in the habit of thinking of the Celts as they were left
by their Saxon and Norman conquerors, a somewhat
backward and relatively thin population in the less ac-
cessible mountain highlands of Scotland and Wales.
But this is only the end of the story which stretches
much further back into the centuries before Christ....
Indeed, the Celtic peoples of the British Isles formed
a part of the great Celtic peoples who occupied and
ruled a large part of Europe before their conquest by
the Romans" (Chadwick, The Celts, Harmondsworth,
Eng., 1970, pp. 7-8). Why would there have been
silence about the Celts, if there has been? Davis
asserts that it is because Celtic civilizations apparently granted women equality under the law, and
Chadwick has claimed that there is archeological evidence in accord with literary evidence suggesting that
"women were accorded parity with men in aristo-
cratic society" (p. 35). There are evidences of
women warriors, not only the great Boudicca of
Britain but also females in charge of military training and priestesses. (Cf. Myles Dillon and Nora K.
Chadwick, The Celtic Realms, London, 1967, pp.
145-6, and Nora Chadwick, Early Brittany, Cardiff,
1969, pp. 31-2.)
    It is somewhat incredible that other ancient
civilizations can be studied by historians of technology,
who generally skip from the classical times to the
Middle Ages, as if nothing intervened, as if many of
the Roman roads and bridges did not likely derive
from the Celts (cf. Chadwick, Early Brittany, p. 44,
for example), as if Celtic transportation, metal-
working, farming by plough, milling by rotary querns
had not existed (cf. the studies by Chadwick and Dillon and Chadwick above), as if many of the sites of
medieval towns and cities had not been Celtic settlements. It is likewise strange that historians of technology can so discuss the building of the pyramids
and obelisks and yet ignore the erection of the great
stone monuments by Celtic peoples in Europe (some
of which are depicted in the Chadwick works, as are
metal artifacts). The medieval cultures are assumed
to be knowledgeable about iron-working, yet little is
mentioned about La Tène Iron Age culture of the
Celts. Though "Celtic culture is the fine flower of
the Iron Age" (Chadwick, The Celts, p. 42), historians of technology have been little interested in
Celtic contributions. "Celtic wheelwrights, carpenters

and blacksmiths cooperated in the production of some
of the most technologically advanced wheeled vehicles
of the ancient world, four-wheeled wagons as well as
the lighter war chariots. The skill of the blacksmith
is seen also in the weapons of the period, particularly the long swords, and in the decorated fire-dogs
which may have graced a chieftain's hearth. In
general, the technological level of La Tène Celts,
with very few exceptions, was equal to, and in some
matters surpassed, that of the Romans" (Ibid., p. 38).
Not only must historians in general and historians of technology and economic historians reassess
Celtic contributions to European history, civilizations
which undoubtedly gave the western world much of
its institutional form; we must also reassess the
ancient world in terms of the linkages between matri-
archy and matrilinear societies, on the one hand, and
economic and technical contributions, on the other.
For example, it is incredible to this historian that
Peter F. Drucker, in "The First Technological Revolution and Its Lessons," Technology and Culture, VII
(1966) 143-151, consistently uses masculine nouns and
pronouns in referring to accomplishments in irriga-
tion when we know that (a) many Near Eastern cultures were matrilinear and (b) often the women were
in charge of agriculture and cultivation. Further,
we know of Egyptian queens, yet discussions of civil
engineering and other technology never examine their
probable involvements in such works. Drucker's
article has only one footnote, so it must be "revealed
truth." Though I am not an expert in ancient history,
I would concur with Dr. K. D. White of the University of Reading, who is--that if scholars are going
to discuss Roman history, some one among them
must ultimately know something about Roman history.
And to reassess ancient technology, we need the combined skills of those who are expert in the history
of technology, ancient history, and women's history.
(Dr. White reviewed The Transformation of the
Roman World: Gibbon's Problem after Two Centuries, ed. by Lynn White, Jr., 1973 in Technology
and Culture, XVI (1975), 481-3.) Also, cf. Autumn Stanley, "Mothers of Invention," SWE Proceedings, 1979.
    From the Brehon Law Tracts (1869), one can see
workmen's compensation, p. 157, and various aspects
of Celtic technology and industry. The Senchus Mor
(Ancient Laws of Ireland), Part II, published London,
1869, pp. liv-lx, indicates the equality of women.

19. The experiment The New Alchemy Institute and its maga-
     zine, Journal of the New Alchemists, along with a
     recent film made concerning this commune were dis-
     cussed by its founder, Nancy Todd, at the conference
     "Lifestyles for Human Growth, Contemporary Chau-
     tauqua," held at the University of Illinois, March
     3-5, 1977. Nancy Todd also presented slides con-
     cerning the technologies of construction used at the
     New Alchemy for the fish ponds, for human habita-
     tion and others.

20. Cowan, "From Virginia Dare to Virgina Slims," p. 42.

21. National Bureau of Economic Research, Inc., Income in
     the United States, Its Amount and Distribution, 1909-
     1919 (New York, 1921), especially pp. 58-60.

22. See, for example, John Kenneth Galbraith, "How the
     Economy Hangs on Her Apron Strings," Ms., II
     (May, 1974), 74-77 and 112. I called this article to
     my husband's attention, and he then used it in one
     of his economics classes; I am indebted to his re-
     ferral of the N.B.E.R. study to me (note 21).

23. Elizabeth Janeway, Man's World, Woman's Place: A
     Study in Social Mythology (New York, 1971), pp. 13-
     21. Janeway draws upon the studies of a number of
     historians such as R. H. Tawney and Carl Briden-
     baugh, as well as works by other respected re-
     searchers.

24. U.S. Women's Bureau, The Occupational Progress of
     Women, 1910-1930 (Washington, D.C., 1933), item
     number 56 in the Soltow and Wery bibliography.

25. Smuts, Women and Work in America, especially pp.
     2-3 and 33-4. Also, see America's Working Women,
     ed. Baxandall et al., pp. 232-40, for the rise of
     women typists and telephone operators.

26. Since both S and C are closely correlated with time,
     the regression coefficients are not highly reliable
     measures of separate influence. We can also set
     the coefficient for S equal to -1 (signifying the as-
     sumption that each decrease of one point in S is
     associated with an increase of one point in C) and
     solve for the other terms: (4) $C = .071 T - 1 S +
     33.6$, yielding $R^2 = .962$.

A simple F-test indicates that the difference
between equations is not statistically significant.
Therefore, we cannot reject the hypothesis that (4)
is the true relationship. Hence the coefficient of
S could be -1, which would be the strongest statement of the relationship between S and C. Note that
equation (1) has a coefficient of S close to -1.

27. W. Elliott Brownlee, "Household Values, Women's Work,
and Economic Growth, 1800-1930," paper delivered at
the Economic History Association, September 15, 1978,
Toronto, Canada, published in the Journal of Economic
History, XXXIX (March, 1979), 199-209.

28. Donald Hoke of the Milwaukee Public Museum delivered a
paper on the history of women and the typewriter at the
Business History Conference, March 2-3, 1979, New
Orleans, published in Business and Economic History,
Ser. 2, VIII (1979). Elyce Rotella has completed a dissertation on Women's Labor Force Participation and
the Growth of Clerical Employment in the United States,
1870-1930 (University of Pennsylvania, 1977), which
contains commentary on the association of female office
help with the typewriter. And James Petersen, in his
paper delivered at the Conference on Critical Issues in
the History of Technology, Roanoke, Virginia, August
17, 1978, on "The Labor Movement" in the session on
"The History of Technology and the Policy Process,"
has noted the comments of typesetters about the entry
of women into typesetting historically.

29. Carol S. Lasser, dissertation in progress at Harvard
on Domestic Service in Early Nineteenth Century
New England.

30. These calculations were made by this author from the
table labeled "Series D 182-232. Major Occupation
Group of the Experienced Civilian Labor Force, by
Sex: 1900 to 1970" on pp. 139-140, U.S. Bureau of
the Census, Historical Statistics of the United States,
Colonial Times to 1970 (Washington, D.C., 1975).

31. Phyllis Chesler, Women and Madness (Garden City,
N.Y., 1972), and Betty Friedan, The Feminine
Mystique (New York, 1963).

32. Scott Donaldson, The Suburban Myth (New York, 1969),
especially the preface and chapter one.

33. Cowan, "Two Washes in the Morning and a Bridge Party at Night: The American Housewife Between the Wars," Women's Studies, III (1976), 151-2.

34. Junius D. Edwards, The Immortal Woodshed (New York, 1955), p. 19.

35. Charles Martin Hall to "Jo," who was likely his sister Julia B. Hall, letter, July 8, 1882, held in the archives of the Aluminum Company of America.

36. Carroll W. Pursell, Jr., "Toys, Technology, and Sex Roles in America, 1920-1940," presented to the Society for the History of Technology, San Francisco, December, 1973, and now appearing in this volume, p. 252.

37. W. Elliott and Mary Brownlee, Women in the American Economy, A Documentary History, 1675 to 1929 (New Haven, Conn., 1976), pp. 2 and 28, respectively for the quotations. Also see W. Elliott Brownlee's paper, delivered at the EHA, cited in note 27.

38. Martha Moore Trescott has written "The Bicycle, A Technical Precursor of the Automobile," Business and Economic History, second series, V (1976), 51-75. This essay does not contain any commentary on women, but the longer manuscript from which it stemmed and which was, in brief, presented to the Business History Conference, Moline, Illinois, March, 1976, does contain some mention of women's roles in the bicycle boom of the nineties. For further information on women, particularly as consumers in this boom, see Robert A. Smith, A Social History of the Bicycle, Its Early Life and Times in America (New York, 1972). Harold F. Williamson, "Mass Production for Mass Consumption," Technology in Western Civilization, I, ed. Kranzberg and Pursell, pp. 686-87, comments on bicycles and briefly on women in relation to the bicycle of the nineties. Of course, even though women did take to bicycles quickly, especially after the safety was introduced in the early nineties, they were often ridiculed. Particularly, those who wore bloomer outfits were criticized by society at large. Discrimination against women and ridicule of them as

bicycle riders can be seen in much of the trade
literature of the day, as well as in the popular
press. At bicycle trade fairs the women were often
ridiculed; The Wheel carried reviews of these fairs
each year during most of the nineties.
Not only as buyers and consumers but also as
producers, women played roles in the booming bicycle business of this period. Trescott has generated
a card file of over 400 bicycle companies in the
1890s, some of which had women officials. These
data can be made available to interested scholars.
Also, as an episode in the history of women's sports,
the bicycle should be studied by women's historians.

39. George H. Daniels, "The Big Questions in the History
of American Technology," Technology and Culture,
XI (1970), 1-21.

40. Melvin Kranzberg, "Technology the Liberator," an address given at the conference "Technology at the
Turning Point: The Rose-Hulman Bicentennial Conference on American Technology--Past, Present,
and Future," April 1-3, 1976. The proceedings of
this conference have been published in W. B. Pickett,
ed., Technology at the Turning Point (San Francisco,
1977).

41. David S. Landes, The Unbound Prometheus, Techno-
logical Change and Industrial Development in Western
Europe from 1750 to the Present (London, 1969).
Ferguson discussed this notion at the SHOT meetings,
Washington, D.C., December, 1969.

42. Thomas C. Guider, "Interconnection and Ideology: The
Case of the Engineers," Society for the History of
Technology, Philadelphia, December 29, 1976.

43. Blau, "Women in the Labor Force," p. 26 of typescript.

44. Robert W. Fogel, "The Economics of Mortality in North
America, 1650-1910," presentation to the Economic
History Workshop, University of Illinois, April 6,
1977.

45. This author checked all of the tables and data listed
under "dentists" and "dental schools" in Historical
Statistics, Colonial Times to 1970, but divisions by
sex were not given.

46. Gerda Lerner, The Woman in American History (Menlo Park, Calif., 1971).

47. Thorstein Veblen, The Theory of the Leisure Class, An Economic Study in the Evolution of Institutions (New York, 1899). Chapters on "Pecuniary Emulation" and "Conspicuous Leisure," note the "barbarian" status of women in our culture.

48. Helen Campbell, "The Working-women of Today," Arena, IV (August, 1891), 329-339, cited in Soltow and Wery's bibliography as item number 14, p. 7.

49. Simone Weil, Notebooks (2 volumes), translated by Arthur Wills (New York, 1956). Cf. also Raymond Sokolov, "Simone Weil: 'The Red Virgin,'" Ms., IV (July, 1975), 99-102.

50. Otto Mayr, "Technology and 'Zeitgeist'" paper delivered at the Conference on Critical Issues in the History of Technology, August 14, 1978, in which he distinguishes four groups involved in technological change: (1) society at large, (2) the technologists, or those who "practice, operate and advance technology," (3) the entrepreneurs (those who control it, or the technocrats), and (4) government.

51. Both David Noble and John Staudenmaier expressed this view, with Noble's main emphasis on the workers and Staudenmaier's on the American Indian, Conference on Critical Issues in the History of Technology.

OVERVIEW

From Virginia Dare to Virginia Slims:
Women and Technology in American Life*

by Ruth Schwartz Cowan

When this topic--women and technology in American
life--was first proposed to me as an appropriate subject for
a bicentennial retrospective I was puzzled by it. Was the
female experience of technological change significantly different from the male experience? Did the introduction of
the railroads, or the invention of the Bessemer process, or
the diffusion of the reaper have a differential impact on the
male and female segments of the population? A careful
reading of most of the available histories of American technology (or of western technology in general, for that matter)
would not lead one to suspect that important differences had
existed. Was my topic perhaps a non-subject? I mulled
over the matter for several months and eventually came to
the conclusion that the absence of a female perspective in
the available histories of technology was a function of the
historians who wrote them and not of historical reality.
There are at least four significant senses in which the relation between women and technology has diverged from that of
men. I shall consider each of them in turn and ask the
reader to understand that what I will say below is intended
in much the same spirit that many of the bicentennial retro-
spectives were intended--to be suggestive, but not definitive.

Women as Bearers and Rearers of Children

Women menstruate, parturate and lactate; men do not.
Therefore any technology which impinges on those processes

*© Ruth Schwartz Cowan, 1977. This essay first appeared in
somewhat different form in W. B. Pickett, ed., Technology
at the Turning Point (San Francisco: San Francisco Press,
1977).

will affect women more than it will affect men. There are many such technologies and some of them have had very long histories: pessaries, sanitary napkins, tampons, various intrauterine devices, childbirth anesthesia, artificial nipples, bottle sterilizers, pasteurized and condensed milks, etc. Psychologists suggest that those three processes are fundamentally important experiences in the psycho-social development of individuals. Thus a reasonable student of the history of technology might be led to suppose that the history of technological intervention with those processes would be known in some detail.

That reasonable student would be wrong, of course. The indices to the standard histories of technology--Singer's, Kranzberg and Pursell's, Daumas's, Giedion's, even Ferguson's bibliography--do not contain a single reference, for example, to such a significant cultural artifact as the baby bottle. Here is a simple implement which, along with its attendant delivery systems(!), has revolutionized a basic biological process, transformed a fundamental human experience for vast numbers of infants and mothers, and served as one of the more controversial exports of western technology to underdeveloped countries--yet it finds no place in our histories of technology.

There are a host of questions which scholars might reasonably ask about the baby bottle. For how long has it been part of western culture? When a mother's milk could not be provided, which classes of people used the bottle and which the wetnurse, and for what reasons? Which was a more crucial determinant for widespread use of the bottle, changes in milk technology or changes in bottle technology? Who marketed the bottles, at what prices, to whom? How did mothers of different social classes and ethnicities react to them? Can the phenomenon of "not enough milk," which was widely reported by American pediatricians and obstetricians in the 1920s and 1930s, be connected with the advent of the safe baby bottle? Which was cause and which effect?[1]

I could go on, using other examples and other questions, but I suspect that my point is clear: the history of the uniquely female technologies is yet to be written, with the single exception of the technologies of contraception.[2] This is also true, incidentally, of the technologies of child-rearing, a process which is not anatomically confined to females but which has been more or less effectively limited to them by the terms of many unspoken social contracts. We

know a great deal more about the bicycle than we do about
the baby carriage, despite the fact that the carriage has had
a more lasting impact on the transport of infants than the
bicycle has directly had on the transport of adults. Although
we recognize the importance of toilet training in personality
formation, we have not the faintest idea whether toilet
training practices have been affected by the various
technologies that impinge upon them: inexpensive absorbent fabrics, upholstered furniture, diaper services, wall-
to-wall carpeting, paper diapers, etc. The crib, the
playpen, the teething ring and the cradle are as much
a part of our culture and our sense of ourselves as
harvesting machines and power looms, yet we know almost nothing of their history.

The history of technology is, of course, a new field
and it is not surprising that its practitioners have ignored
many of the female technologies. We do not usually think
of women as bearers of technological change nor do we think
of the home as a technological locale (in part because women
reside there). Both of these common assumptions are incorrect; Adam knew that, but his descendants have forgotten it.

Women as Workers

Women have been part of the market economy of this
country from its earliest days. In the colonial period they
tended cows, delivered babies, kept taverns, published newspapers and stitched fancy clothes, among other things. During industrialization they tended looms, folded paper bags,
packed cigars, helped with harvests, washed laundry, and
stitched fancy clothes, among other things. With the advent
of automation they punch cards, handle switchboards, pack
cookies, teach school, tend the sick and stitch fancy clothes,
again among other things. All along they have been paid for
their work, sometimes in land, sometimes in produce, sometimes in cash.

But women workers are different from men workers,
and the differences are crucial, for the women themselves
and for any analysis of the relation between women and the
American technological order. The economic facts of life
for women are almost on a deterministic par with the ana-
tomic facts of life; they are so pervasive over time and
place as to be almost universal truisms. There are three

of them: (1) when doing the same work women are almost always paid less than men; (2) considered in the aggregate, women rarely do the same work as men (jobs are sex-typed); (3) women almost always consider themselves, and are considered by others, to be transient participants in the work force. [3]

These characteristics of women as workers predate industrialization; they were economic facts of life even before our economy was dominated by cash. Sex-typing of jobs occurred in the earliest Jamestown settlements, even before the household economy had completely replaced the communal economy; unmarried or poor women worked as laundresses in return for a portion from the communal store; men who were not entitled to grants of land worked as cooks and bakers. [4] Unequal pay for equal work was also characteristic of the early settlements; adventurers who came to settle in Maryland were allotted 100 acres of land for every man-servant they brought with them and 60 acres of land for every womanservant; unmarried free men in Salem and Plymouth were given allotments of land when they requested them, but after the first few years of settlement, unmarried women were not; the first American effort to obtain equal economic rights for women may well have been the request made in 1619 by the Virginia House of Burgesses that husbands and wives be granted equal shares of land on the grounds that the work of each was equally crucial to the establishment of a plantation. [5] That women were regarded as transitory members of the work force even then is shown by many things: for example, the fact that when girl children were "put out" for indenture or apprenticeship the persons who received their work were rarely required to teach them a trade, [6] or the fact that women who owned and operated businesses in the colonial period were almost always widows of the men who had first established the business, who consciously advertised themselves as worthy of patronage on those grounds alone. [7] Parents of daughters did not expect that their girl children would need to know any occupation other than housework; young women expected that they might need to support themselves while unmarried but that gainful employment would become unnecessary and undesirable after marriage; married women expected not to be gainfully employed unless their husbands died or were disabled.

So it was, and so it continues to be--despite industrialization, unionization and automation. The Equal Pay Act of 1964 attempted to legislate equal pay for equal work

for women but in 1973 it was still true that women
were earning from 37.8 to 63.6 per cent of what men
in the same job classifications were earning. [8] Power
technologies have eased and simplified thousands of jobs,
yet the labor market is still dominated by sex-typed occupations, despite the fact that the worker's "strength" is no
longer a relevant criterion. In the garment industry in New
York, for example, men cut and women sew. Thus when a
manufacturer goes into the labor market to find employees,
he or she enters one labor market, with its own price
structure and its own supply-demand pattern, if searching
for a skilled cutter--and a different labor market, with a
price structure and a supply-demand pattern all its own, if
looking for a skilled sewing machine operator. [9] A fairly
sophisticated statistical analysis of sex-typing in the labor
market has demonstrated that although some job classifications shifted from being male-dominated to being female-
dominated between 1900 and 1960 (ironically, none has gone
the other way), the total amount of sex-typing has not changed
appreciably. In 1900, 66 per cent of all employed women
would have had to shift their jobs into male-dominated fields
in order for the distribution of women and men in all fields
to resemble chance; in 1960 that figure was 68.4 per cent.
(The corresponding figure, incidentally, for racial typing of
jobs in 1960 was 46.8 per cent). [10]

And of course it is still true, as it was in colonial
days, that women are not regarded, by themselves or by
others, as permanent members of the work force. Many
young women do not invest in expensive training for themselves because they anticipate leaving the work force when
they marry and have children. Employers are equally un-
willing to invest in training women because they anticipate
the very same thing--and with some statistical basis for their
suspicion; the labor force participation rates of females
typically has fallen off sharply between the ages of 18 and
25, the years when most women marry and begin their
families. [11] The cumulative result of these attitudes is that
women place themselves in fairly unskilled, unresponsible
and therefore lower-paying positions and employers are content to have them remain there.

These three characteristics of women as workers--
the fact that they work for less, that many jobs are not open
to them because of sex typing, and that they are transient
members of the work force (and therefore difficult to organize and unionize)--should be of signal importance in any

discussion of rates of technological change, but they are
rarely considered in that context. We are accustomed to
thinking about the price and availability of labor as one of
the key determinants of rates of change in any given industry or any given locality, but we are not accustomed to
thinking of the price and availability of labor as determined
by the sex of the laborers. The ways in which the sex of
workers interacts with technological change can be illustrated
by two somewhat different cases.

The first is the cigar industry in the second half of
the nineteenth century, a case in which technological change
was accelerated by the availability of female workers.[12]
During the middle decades of the century cigar-making was
localized in factories but the product was entirely handmade
by skilled male workers, most of them Spanish, Cuban and
German. In 1869 the New York cigarmakers went on strike
and in retaliation, several manufacturers arranged for the
immigration of Bohemian women who worked in the cigar
trade in their native land. These women were not as skilled
as the men they replaced; they used a simple molding tool
to shape the cigar. They also were accustomed to working
at home, which meant that they were amenable to the tenement system of manufacture, which was much cheaper for
the employers. The women were effective in breaking the
strike. In subsequent years more women cigarmakers immi-
grated as the cigar trade in Bohemia was disrupted by the
Franco-Prussian War. There were other male cigarmakers'
strikes in New York and elsewhere during the 1870s and
1880s, the net effect of which was that some manufacturers
converted entirely to the tenement system and others were
induced to try some simple pieces of machinery (also of
European origin) which could be operated by women. The
women were laborers of choice because they knew the cigar
trade, yet were willing to work for less and had not been
organized. As one New York cigar manufacturer put it in
1895: "... the handwork has almost entirely disappeared.
The suction tables, which are in reality nothing else than
wrapper cutting machines, are used as price cutters. More
so, because there are only girls employed on them."[13]

A somewhat contrary case is that of the ladies garment industry in the twentieth century; here technological
change seems to have been slowed by the presence of female
workers.[14] Since the time that sewing machines were
initially hooked up to central power supplies (steam or electricity) in the last quarter of the nineteenth century, there

has been little technological change in the sewing process, despite the fact that the industry is highly competitive and despite substantial changes in the technology of the processes that are auxiliary to sewing--namely cutting and pressing. The sewing process could potentially be automated but there appears to be little incentive for manufacturers to do this, partly because the expense would be very great and many of the companies are very small. Yet another reason stems from the fact that sewers are women and sewing is work that women from traditional cultures like to do. The ladies garment industry has been populated by successive waves of fairly skilled immigrant women of various sorts: American farm girls who came to the cities to escape rural life in the middle of the last century, then immigrant women from Italy and eastern Europe, then black women from the south, then Puerto-Rican women, and now Chinese women. Although these women are skilled and although their trade has been unionized successfully for many years, the wages paid to sewing machine operators are, as one would expect, significantly lower than the wages paid to other skilled machine operators. As a consequence, the technology of sewing has remained fairly static.

There is yet another sense in which the characteristics of women as workers have interacted with the technological order in this country--and here we confront one of the most firmly grounded shibboleths about the relation between women's work and technology. It is true that there has been a vast increase in the number of women in the work force in the past century and during this time some occupations (such as clerical work) have almost completely changed sex. It is also true that during the same period power-driven machinery has entered many fields, requiring much less human energy to do work that was once hard to do. Acknowledging that these facts are true, historians and others have concluded that technological change has drawn women into the work force by opening fields of work that were previously closed to them because of the physical strength required to do the work. That conclusion appears to be almost entirely unwarranted. Women have replaced men in several occupations in which hard physical labor was not required before industrialization (for example, cigar making); they have replaced men in some occupations in which no significant technological change occurred (for example, school teaching); and they have replaced men in some occupations in which technological change made no difference to the physical labor involved (for example, bookkeeping). In all of these cases, the crucial factor is not physical labor but price; women replaced men because they worked for less.[15]

Alternatively, there are many trades in which work has been transformed by the introduction of new machines but in which women have not replaced men. Typesetting is a perfect example.[16] From the colonial period to the present there have always been a few women typesetters, but they have worked in the smallest shops, often shops that were family-owned. Typesetting was generally an apprenticed trade in the nineteenth century and women were not set to apprenticeships. In any event it was widely believed that women could not do the work of typesetting efficiently because they were not able to carry the heavy type cases from the composing tables to the press. Women typesetters were occasionally used to break strikes or to cut wages, a practice which did not endear them to the typesetters' unions which were formed during the early decades of the century. In 1887 the linotype machine was introduced and after that time the work of typesetting was not terribly much more difficult than the work of typewriting. Various modifications of that machine, and the more recent introduction of photographic processes, have made the work easier still--but men dominate the trade. The reasons for this are several: after women were admitted to the typesetters' unions, in the latter part of the nineteenth century, they had to agree to work for scale, which meant, of course, that employers had little interest in hiring them; following this, the advent of protective labor laws meant that night work for women was very carefully regulated, which made it unfeasible for women to become typesetters since so much of the work is on newspapers. Thus technological change has been, at best, a mixed blessing for women. More jobs are open to them that they are fit (either biologically or socially) to perform, but many of those jobs are at the very lowest skill and salary levels and are likely to remain that way as long as women are willing, for whatever reasons, to work for less than men and to let themselves be treated as marginal members of the labor force.[17]

Women as Homemakers

Both men and women live in homes, but only women have their "place" there--and this is another one of those salient facts about women's lives which make their interaction with technology somewhat different from men's. The homes in which we live, the household implements with which we work, and the ways in which that work is organized have changed greatly over the years, but the character of that change and its impact upon the people who work in homes (predominantly women) has proved very difficult to gauge.

Some tasks have disappeared (e.g., beating rugs) but other tasks have replaced them (e.g., waxing linoleum floors). Some tasks are easier (e.g., laundering) but are done much more frequently; it takes less time and effort to wash and iron a sheet than it once did, but there are now vastly more sheets to be washed in each household each week. Some tasks have become demonstrably more time-consuming and arduous over the years: shopping, for example.[18] Less work needs to be done at home because so many aspects of home production have been transferred into the market place (e.g., canning vegetables), but there are now fewer hands to do the work as there are fewer servants, fewer unmarried females living at home and fewer children per family. An equivocal picture at best.[19]

But one point is worth making. Despite all the changes that have been wrought in housework, and there have been many, the household has resisted industrialization with greater success than any other productive locale in our culture.[20] The work of men has become centralized, but the work of women remains decentralized. Several million American women cook supper each night in several million separate homes over several million separate stoves-- a specter which should be sufficient to drive any rational technocrat into the loony bin, but which does not do so for reasons I will discuss in a moment. Out there in the land of household work there are small industrial plants which sit idle for the better part of every working day; there are expensive pieces of highly mechanized equipment which only get used once or twice a month; there are consumption units which weekly trundle out to their markets to buy eight ounces of this non-perishable product and twelve ounces of that one. There are also workers who do not have job descriptions, time clocks, or even paychecks.[21] Cottage industry is alive and well and living in suburbia.

Why? There is no simple answer to that question but I would like to attempt a rough list of what some of the components of the answer might be, presenting them in no particular order and with no pretense of knowing the relative weight which should be attached to each, or the relative

likelihood that some are causes and others effects. To start
with, since the middle of the nineteenth century Americans
have idealized the household as a place where men could retreat from the technological order: a retreat, by definition,
should not possess the characteristics that one is trying to
avoid. Increased efficiency and modernity in the home have
occasionally been advocated by domestic reformers, most of
whom have been women (Catherine Beecher, Ellen Swallow
Richards, Charlotte Perkins Gilman, Christine Frederick and
Lillian Gilbreth immediately come to mind), but the general
public has been hostile to certain crucial concomitants of
those ideas. [22] The farm kitchen has been the American
mythic dream, not the cafeteria. In some households the
latest and showiest equipment is purchased in order to demonstrate status, not efficiency; in such cases the housebound
housewife is as much proof of status as the microwave oven
that she operates. In other households status is not the issue, but modern equipment is used to free the housewife for
labor which is not currently technology-intensive (stripping
furniture, planting vegetable gardens, chauffering children),
and the end result is far from an increase in overall efficiency. Except for a very brief period in the 1920s, "Early
American" has been, far and away, the most popular decor
for American kitchens; our ambivalence on the issue of efficiency in the home is nowhere better symbolized than when
a dishwasher is built into a "rustic" cabinet or a refrigerator
is faced with plastic "wood" paneling. For long periods of
time, on either side of the industrial revolution in housework
(which can be roughly said to have occurred in the first
three decades of this century), the maintenance of a fundamentally inefficient mode of household operation, requiring
the full attention of the housewife for the better part of every
single day, has been a crucial part of the symbolic quality
of the individual American home.

Connected with this is the fact that also since the
middle of the nineteenth century most Americans have regarded communalization of households as socialistic and
therefore un-American. There have been repeated attempts
at communalizing some of the household functions, especially
during the first two decades of this century--communal can-
neries, laundries, kitchens, even nursery schools appeared
in many communities--but they have all failed for want of a
supportive community attitude. [23]

The implements invented or developed for the home
have very special features which may set them apart from

other implements.   Many of them were initially developed,
for example, not for home use but for commercial use: the
automatic washing machine, the vacuum cleaner, the small
electric motor and the refrigerator, for example.[24]   Most
of them were not developed by persons intimately connected
with the work involved; inventors tend to be men and homemakers tend to be women.   On top of this, many of the
implements were marketed through the use of selling techniques that also had little relation to the work performed.
These three factors lead to the hypothesis that the implements which have transformed housework may not have been
the implements that housewives would have developed had
they had control of the processes of innovation.

Thus, for reasons which may have been alternately
economic, ideological and structural, there was very little
chance that American homes would become part of the industrial order in the same sense that American business has,
because very few Americans, powerful or not powerful, have
wished it so--and the ones who have wished it so have not
been numerous enough or powerful enough to make a difference.

Women as Anti-Technocrats

Which brings me to a final and somewhat related
point.   For the better part of its cultural life the United
States has been idealized as the land of practicality, the
land of know-how, the land of "Yankee ingenuity."   No
country on earth has been so much in the sway of technological order nor so proud of its involvement in it.   Doctors
and engineers are central to our cultures; poets and artists
live on the fringes.

If practicality and know-how are signatures of the
true American, then we have been systematically training
slightly more than half our population to be un-American.
I speak, of course, of women.   While we socialize our men
to aspire to feats of mastery, we socialize our women to
aspire to feats of submission.   Men are hard; women are
soft.   Men are meant to conquer nature; women are meant
to commune with it.   Men are rational, women irrational;
men are practical, women impractical.   Boys play with
blocks; girls play with dolls.   Men build; women inhabit.
Men are active; women are passive.   Men are good at
mathematics; women are good at literature.   If something

is broken, daddy will fix it. If feelings are hurt, mommy will salve them. We have trained our women to opt out of the technological order as much as we have trained our men to opt into it.

This is probably just as much true today as it was in the heyday of the archetypically passive, romantic Victorian female. An interesting survey of American college girls' attitudes towards science and technology in the 1960s revealed that the girls were planning careers but that they could not assimilate the notion of becoming engineers and-- this is equally revealing--that there was no single occupation that they thought their male contemporaries and their parents would be less pleased to have them pursue. [25]

Thus women who might wish to become engineers or inventors or mechanics or jackhammer operators would have to suppress some deeply ingrained notions about their own sexual identity in order to fulfill their wishes. Very few people have ever had the courage to take up such a fight. It is no wonder that women have played minor roles in creating technological change; in fact, it is a wonder that there have been any female engineers and inventors at all.

Conversely, it may be true that the recent upsurge in "anti-science" and "anti-technology" attitudes may be correlated very strongly with the concurrent upsurge in women's political consciousness. This is not to say that all of the voices that have been raised against the SST and atomic power plants and experimentation on animals have been female--only that a surprisingly large number of them have been (think of Rachel Carson and Frances Simring Robinson). Ann Douglas has recently written a complex analysis of the "feminization" of American culture in the nineteenth century in which she suggests that the "tough minded" theological attitudes that had served as cornerstones of American ideology in the seventeenth and eighteenth centuries were watered down and whittled away in the nineteenth by several generations of educated and literary women working in concert with similar generations of liberal male theologians. [26] Both groups, she argues, realized that they were excluded from the burgeoning capitalist economy that the older theology had produced; they resented this exclusion and so fought against the economy and the theology together.

The temptation to push Douglas's analysis into the twentieth century is almost irresistible. If we are experiencing a similar "feminization" of American culture today, it is the "tough minded" ideology of the scientific-technocratic state, that is the focus of current animus. Women have traditionally operated on the fringes of that state, so it is not surprising that they should resent it and, when given the opportunity, fight against it.

Women have experienced science and technology as consumers, not as producers--and consumers, as every marketing expert knows, are an infuriatingly fickle population. Trained to think of themselves as the possessors of subjectivity, women can hardly be expected to show much allegiance to the flag of objectivity. As more and more women begin to play active and powerful roles in our political and economic life, we may be surprised to discover the behavioral concomitants of the unspoken hostility to science and technology that they are carrying with them into the political arena.

Notes

1. The history of nursing practises in early modern Europe is surveyed in Edward Shorter, The Making of the Modern Family (New York, 1975), Ch. 5, but developments since the appearance of pasteurized milk and sterile (or sterilizable) bottles are not considered.

2. Norman E. Hines, A Medical History of Contraception (Baltimore, 1936) does not cover more recent developments. Linda Gordon, Woman's Body Woman's Right: A Social History of Birth Control in America (New York, 1976), focuses on ideas about birth control, but not on the devices themselves.

3. On the economics of workers, see Robert W. Smuts, Women and Work in America (New York, 1959) and Juanita Kreps, Sex in the Marketplace: American Women at Work (Baltimore, 1971).

4. Julia Cherry Spruill, Women's Life and Work in the Southern Colonies (Chapel Hill, North Carolina, 1938), p. 6.

5. Spruill, pp. 9-11; also Edith Abbott, Women in Industry: A Study in American History (New York, 1910), p. 11.

6. Abbott, pp. 30-32.

7. Spruill, passim.

8. United States Department of Labor, 1975 Handbook on Women Workers, Bulletin 297 (Washington, D.C., 1975), p. 130.

9. On sex typing of jobs, see Valerie Kincaid Oppenheimer, The Female Labor Force in the United States, Population Monograph Series, No. 5 (Berkeley, California, 1970).

10. Edward Gross, "Plus ca change ... The Sexual Structure of Occupations Over Time," Social Problems, 16, 1968, 202.

11. Kreps, pp. 28-29.

12. Abbott, Ch. 9.

13. Abbott, p. 263. Patricia Cooper, a graduate student in history, at University of Maryland, College Park, will soon complete a dissertation on tobacco workers which extends and reinforces these conclusions.

14. This account is based on Elizabeth Faulkner Baker, Technology and Women's Work (New York, 1954), Ch. 15.

15. See Baker, Chs. 13, 17.

16. Baker, pp. 170-177; Abbott, Ch. 11.

17. Interest in women workers has revived in the last few years: see articles by Susan Levine on carpet workers and Daryl Hafter on drawgirls in this volume, as well as Judith A. McGaw, The Sources and Impact of Mechanization: The Berkshire County, Massachusetts Paper Industry, 1801-1885, as a Case Study (Ph.D. dissertation, New York University, 1977) and also the McGaw essay in this volume; Rosalyn Baxandall, et al. eds., America's Working Women: A Documentary History (New York, 1976); and Barbara Mayer Wertheimer, We Were There: American Women Who Worked, (New York, 1977).

18. On time spent in housework, see JoAnn Vanek, Keeping Busy: Time Spent in Housework, United States, 1920-1970 (Ph. D. Dissertation, University of Michigan, 1973).

19. For an extended discussion of this topic see my papers, "The 'Industrial Revolution' in the Home: Household Technology and Social Change in the 20th Century," Technology and Culture, 17 (1976), 1-22 (reprinted, this volume), and "Two Washes in the Morning, and a Bridge Party at Night: The American Housewife Between the Wars," Women's Studies, 3 (1976) 147-172.

20. On this point see Allison Ravetz, "Modern Technology and an Ancient Occupation: Housework in Present Day Society" Technology and Culture 6 (1965) 256-260. Also, for a general history of housework, see Ann Oakley, Woman's Work, The Housewife Past and Present (New York, 1974).

21. For an extended discussion of the sociological meaning of this phenomenon, see Ann Oakley, The Sociology of Housework (London, 1974).

22. See Kathryn Kish Sklar, Catherine Beecher: A Study in American Domesticity (New Haven, 1973); Caroline Hunt, The Life of Ellen H. Richards (Boston, 1912); and Waida Gerhardt, "The Pros and Cons of Efficiency in the Household," J. Home Eco., 18 (1928) 337-339.

23. Victor Papanek and James Hennessey speculate about how various implements would have to be redesigned for communal ownership in How Things Don't Work (New York, 1977), Ch. 2.

24. Siegfried Giedion, Mechanization Takes Command (New York, Oxford University Press, 1948), pp. 556-606.

25. Alice Rossi, "Barriers to the Career Choice of Engineering, Medicine or Science Among American Women," in Jacqueline Mattfeld and Carol VanAken, eds., Women and the Scientific Professions (Cambridge, Mass.: M.I.T. Press, 1965), pp. 51-127.

26. Ann Douglas, The Feminization of American Culture (New York, 1977).

PART I

Women as Active Participants
in Technological Change

INTRODUCTION TO PART I(A)

Women Operatives and Workers in Industry

Two of the three essays in this section deal with women workers in textile operations other than the cotton industry, which has already been treated in some detail by historians of women in industry. Daryl M. Hafter, a professor of history at Eastern Michigan University, sheds light on the "decline of the drawgirl" in the French brocade industry of the eighteenth century. As mechanization of weaving became more entrenched with the coming of the Jacquard loom and other innovations, the drawgirls, or tireuses, were displaced. In this displacement process, Hafter has found much anti-female propaganda directed at these women workers. Next, we move to the United States in the late nineteenth century and there view carpet manufacture. Susan Levine has found that with the rise of power looms, female weavers actually displaced skilled male weavers, primarily because women worked for significantly lower wages. Hafter has also noted in closing that after the 1790s, French women increasingly became weavers on the newer looms. In each case, the new technology helped break the old guild and artisan craft traditions, thus allowing more women to become weavers. Then, while mechanization displaced some women workers such as the drawgirls, it was nevertheless accompanied by an influx of women into jobs which had previously been male-dominated.

Judith A. McGaw, a professor at the University of Oklahoma, has found, on the other hand, that mechanization in the Massachusetts paper industry in the nineteenth century did not greatly affect sex segregation on the job. Women essentially continued doing the same jobs before and after the introduction of various machines which were typically tended by men. Women's work continued to be unskilled but, because it was not generally involved with machines, the working

47

conditions of women improved relative to the men's. McGaw
also views the ages and marital and economic status of women
working in paper mills, 1820-1885, and finds some significant
departures from the "consensus" mill girl of the cotton industry.

As stated in the general introduction, these essays
tend to underscore Ruth Cowan's overview analysis in various ways. In particular, women workers have typically
worked for lower wages than men and have performed different tasks on the job than men. Interestingly, mechanization helped differentiate the jobs done by male and female
weavers, for with machines employers could substitute the
less-skilled labor of women for that of the skilled male
weavers, thus somewhat altering the nature of the work.

Sex-typing of work, of course, has existed before the
introduction of machines and afterward. In some cases (as
noted by Levine) mechanization went hand-in-glove with the
conversion of previously male jobs to predominantly female
work. In some cases, machines displaced female workers
altogether (as noted by Hafter), and in some cases the introduction of machines did not essentially change the sex-typing
of work (noted by McGaw).

One interesting point made by McGaw serves as an
exception to Cowan's characterization of women as transient
workers. In the paper mills of Berkshire County, Massachusetts in the nineteenth century, McGaw has found that a
large proportion of the female workers were married and
long-term employees.

1. THE PROGRAMMED BROCADE LOOM AND THE "DECLINE OF THE DRAWGIRL"

by Daryl M. Hafter*

I

Recent historians of technology have focused attention on the dynamics of invention in the hope of formulating a philosophy of mechanical development. In the quest to isolate the general characteristics of invention, certain principles have been offered as constants in the process. An important debate has arisen between those who see the fundamental nature of invention proceeding from the material qualities of the crafts, and those who lay more emphasis on the social determinants of creativity. The first view claims there is an innate process by which industrial tasks call forth machines to accomplish them more efficiently, and the early machines then provide an inborn compulsion for further improvements. While proponents of the latter view acknowledge that external factors influence mechanical developments, they are fundamentally in disagreement with those who view invention as a rational response to economic conditions. This study suggests that neither view is universally true and thus proposes a middle way between the two schools, both in terms of isolating what motivates the making of inventions and in proposing that the path to new uses of machines is much less direct than we would like to think. [1]

The development of a self-acting loom capable of making patterned cloth is a case study in these two principles. The loom itself offered an important problem that intrigued inventors and the desire to improve it functioned

*The author gratefully acknowledges a grant from the American Philosophical Society (Johnson Fund) which supported the research for this study. Thanks are also due Eugene S. Ferguson, Rita Adrosko, Melvin Kranzberg, Mary Robischon and Fred Anderson for their helpful comments.

as a kind of determinism, irrespective of external setting.
The lure to make improvements grew from the early looms'
nature: made of wood, which was easy for an amateur to
work, situated in home workshops, grouped usually in sets
of four so that unmounting one would not bring the business
to a halt, hand looms were objects, par excellence, of
mechanical tinkering. Small improvements were an immediate aid to manufacture, while the possibility of revolutionary improvements in loom technology lent glamor to
every new scheme of cords and shuttle.

Set against the economic fluctuations of eighteenth-
century French industry, the persistent quest for an automatic brocade loom demonstrates how tenuous the notion of
economic rationality may be. This useful ideal overlooks
the fact that society only vaguely follows actions that would
efficiently pursue its economic well-being. While the success of French silk brocades through the first half of the
eighteenth century provided a rational basis for developing
a more efficient brocade loom, the decline of brocades after
1780 was not matched by falling interest in the invention.
Rather, inventors seized upon new rationales to justify continued support of their efforts. They took full advantage of
the French obsession with England's apparent technical superiority to advocate aid to inventors as a patriotic dona-
tion. In addition, they denigrated the workers who would be
replaced, in order to make the new devices appear necessary. Thus Jacquard's final breakthrough in 1807 came not
in response to general economic needs, but as a technical
tour de force. Only afterward, when the Jacquard apparatus
was generalized and its mechanism improved in the nineteenth century, did its industrial uses breed a self-generating
need for the invention.

II

The drawloom itself had been brought to France in
the sixteenth century from Italy, then the silk center of
Europe. Important mechanical capabilities of the loom were
subsequently developed by the French. Attempts to make a
patterning process that the weaver could manage alone began
early in the eighteenth century. In the grande tire looms,
which made the fancy brocades that won Lyons a world reputation, the design was programmed by means of a semple,
the group of cords that hung alongside the loom. From the
first, these cords were worked by girls or boys who formed

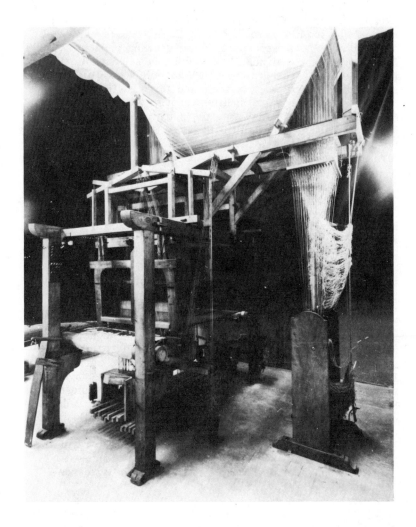

A drawloom following the design of Philippe De Lasalle, showing a semple with the type of wooden grill he used. In this version the semple is still a non-detachable part of the loom. Reconstructed by the Ecole de Tissage, Lyons, France. (Courtesy: Musée Historique des Tissus.)

part of the auxiliary trades that prepared the thread and
readied the loom for weaving. In Paris, Tours and other
silk centers, boys were used in the drawing capacity.
Lyons was somewhat unique in relying upon young women
for this work. The drawgirls had to pull a cord on the
semple forward to activate the warp threads; then the weaver
could pass one or more shuttles through the warp shed to
form the design. To advance the pattern the drawgirl would
then grasp the semple with both hands and tug it down briskly,
bringing into reach the next series of looped cords that determined the subsequent line of the pattern.

This laborious method of advancing the pattern had
obvious defects. It was slow, halting and subject to error.
It depended on the perfect synchronization of the drawgirl
and the weaver. The weaver could not work with continuous
and smooth action. Until new devices like the Garon lever
(or fork) began to be used, drawing down the cords tended to
make patterns irregular. Of equal nuisance value was the
investment of time and money needed to prepare the semple
before weaving could begin. Before the invention of a re-
movable semple in 1775 by the designer Philippe De Lasalle,
the cords of the semple had to be mounted onto the loom itself, immobilizing it for weeks at a time. The most skilled
and costly of the auxiliary workers then 'read' each line of
warp from the designer's plan, and prepared a series of
loops to govern each change of color. This procedure kept
the potential production down, since an entire semple hanging
from ceiling to floor governed only about a foot of finished
cloth. Weavers who completed the semple on one loom,
coming to the end of the design, had to move to another
loom while the first was remounted. Since guild regulations
permitted a master weaver to have only four looms at a
time, the output could not be increased or the price decreased.[2]

These obstacles inspired the search for improvements.
Attention focused on the invention of a brocade loom that
could be worked by one weaver alone. Eliminating the drawgirls, or tireuses, from the weaving process became almost
an obsession among silk producers. The claim to have
made a machine that succeeded even partially in this attempt
virtually guaranteed a government subsidy. Even though few
of the subsidies were very large in themselves, the manufacturer's precarious financial affairs made them desirable.
Indeed Brisson, the disgruntled inspector of manufactures
who supplied Diderot with information on the industry, claimed

Engraving of a drawloom for making fancy silk brocades.
The cordage stretching from ceiling to floor "R" is the
semple which drawgirls pull to determine the pattern.
"Soierie, Etoffes Brochées, Elevation Perspective du Métier
pour fabriquer les Etoffes Brochées garni de tous les Cordages
et agrès." From Encyclopédie. Recueil de planches, vol. 11
(Paris: Briasson, 1772), pl. 60. Photo courtesy Winterthur
Museum Libraries.

---

that the practice had become fraudulent, with friendly manufacturers swearing to the efficiency of looms that had serious
defects.[3]

    With hindsight we view the series of loom improve-
ments as precursors of the successful Jacquard loom which
finally permitted the weaver to work alone. Several paths
to this goal were pursued. The most significant was begun
in 1725 by a weaver named Basile Bouchon, who substituted
a roll of pierced paper for the semple. Bouchon's invention

was improved by Falcon, who replaced the roll of paper with
a series of pierced, stiff cards laced together. Neither permitted the weaver autonomy, though, since they made use
of helpers to manage the pierced papers. Nevertheless Falcon's invention was considered of great practical value and
weavers who agreed to buy such a loom were permitted to
add a fifth loom to their workshops. The novel features of
these looms, the use of pierced paper or cards instead of
the semple, became the hallmark of the Jacquard loom. [4]
But other features also received innovative treatment by inventors.

    The cylinder was the object of the greatest number
of usages. In 1740 an inhabitant of Nimes, Régnier, devised a cylinder with needles attached to the cords of the
semple. Every rotation governed a pick of weft and the
longitudinal sections of the cylinder could be separated.
The fact that the diameter of the cylinder corresponded to
the dimensions of the cloth design limited its usefulness. [5]
The same problem limited the most famous of these innovations using the cylinder, that of Jacques Vaucanson, the
mechanical genius who won fame with his automated figu-
rines. In 1745 the Mercure de France credited him with
inventing a loom activated by means of a windlass. Instead
of tailcords, Vaucanson's loom was made self-acting by
means of a large cylinder around which he placed a band of
pierced paper that determined the design. The back and
forth motion of this "drum" was also linked to a device for
regulating the thickness of the warp and the force of the
beater so that the cloth could be woven evenly on both ends.[6]
While Vaucanson's admirers projected the use of this remarkable machine for brocaded fabrics, when it was praised
in 1745 it had successfully produced only plain silks like
taffeta, serge and satin. [7]

    There is some doubt whether Vaucanson's invention
became industrially practical but it did inspire a number of
fabric manufacturers to adapt the cylinder to functioning
looms. Jacques and Nicolas Currat seem to have been the
first to apply for a grant using this principle. In 1754 they
claimed that they had "discovered how to activate the loom
by a single person with as much facility and skill as if there
were two, which seems impossible if one thinks of the principle of the machine."[8] They claimed that diminishing the
weights enabled them to form straighter lines. The design
was made by means of a drum barrel composed of two
hundred and fifty wooden spokes and a disc with an equal

number of teeth. They devised a cam to activate the griffe, a moving member that permitted the cylinder to rotate by means of its back and forth motion. The addition of the cam was as important as the pierced cardboard in creating a one-person brocade loom and later for converting the loom into a powered machine. [9]

The Currat brothers' device inspired a silk manufacturer named Flachat, who replaced the semple with a wheel in 1760. His invention was activated by the weaver's foot and worked by means of a series of spokes which caught the cords that worked the design. Ten years later, another silk manufacturer, Maugis, devised a loom that functioned on similar principles. In 1770, when the Grande Fabrique of Lyons was in full power, an inquiry was made concerning those looms at work which successfully functioned without the drawgirl. Of seven that are described, only two, those of Maugis and a M. Aulard, worked well by means of the weaver alone. [10]

III

Although an industrially popular loom without drawgirls had not been achieved in fifty years, the French were not ready to give up. The desire to emulate the English industrial miracle lent fresh stimulus to inventors. French technicians were aware of English improvements in the weaving and spinning processes and tried to view the novelties in England or lure their inventors and workmen away. The cotton industry had been the beneficiary of most of these inventions, from John Kay's flying shuttle to Crompton's mule and Cartwright's power loom. Although many of the devices were originally aimed at improving the woolen industry, they resulted in the installation of 1,400 viable automatic looms by 1813, turning out English cottons. [11] These improvements helped to facilitate an extraordinary development of cotton cloth which far surpassed the French production. [12] Although the French borrowed elements of these machines, they were responsible for the major innovations in silk looms, just as they dominated the European silk industry.

The French emphasis on luxury fabric rather than goods for the ordinary market had ideological implications that shaped technology. While in England there was strong commitment to labor-saving devices, in France the mercantilist notion that work must be found for the largest number

of hands prevailed. As late as 1784, brocade looms with
drawgirls were praised over looms making plain cloth because the brocade looms employed twice as many workers.
From this viewpoint, it was "the benefit of labor which remains in the towns when the products have left that is the
real product of the manufactures." The purpose of investment should be "to attract and fix within our walls many useful subjects."[13] The persistence of this tenet, that industry
should be the means of encouraging population growth and
that the most valuable earnings were labor fees, opposed
the inventors' goal. An evident paradox emerged as the inventors worked toward eliminating excess workers and argued
in favor of efficient machines in their particular crafts, without explicitly attacking the traditional ideal that industry
should continuously expand employment. Supporters of the
loom replacing drawgirls had to rely on a variety of practi-
cal-sounding arguments to justify the elimination of work.

At times the inventors complained of the scarcity of
drawgirls. Attracted mainly from the deprived regions of
the Savoy and the provinces surrounding Lyons, the tireuses
formed part of the large proletariat employed in the Grande
Fabrique. Lyonnais women were warned off the auxiliary
trades because of their harsh working conditions, but the
manufacturers sometimes asserted that the potential drawgirls from Savoy also disdained the work. Manufacturers
then found it hard to fill their orders and had to recruit the
drawgirls from their native regions. On the other hand, the
inventors were also concerned to explain what would become
of the ten thousand or so drawgirls currently employed.[14]

The Currat brothers thought it necessary to assure
officials that their invention would cause the girls to diversify,
not deprive them of work. "On the contrary," they wrote,
"the number of auxiliary workers will increase because those
who now pull the cords will occupy themselves with other
necessary work to prepare the silks for manufacture like
reeling it off cocoons and preparing the loom. These tasks
always lack workers, especially the reeling. And the girls
who are in Savoy where the largest part know that their colleagues have jobs other than drawing, will come with pleasure
and occupy themselves with preparing the silks. Instead of
being hired right away as soon as they arrive in Lyons by
individual masters, and binding themselves to being drawgirls, they will be instructed in other work with which they
can occupy themselves."[15] Later commentators appealed to
the prejudice that the cities were depopulating the countryside.

They suggested that the former drawgirls would be freed to
return to their homes in rural areas and undertake farming.
This advice ignored the fact that the girls had emigrated
from marginal farms which could not support them.

Jobs for women were considered to be a necessary
measure to prevent prostitution; therefore an inventor who
promoted labor-saving devices had a moral obligation to suggest other employment. Another way to show concern for
human needs was to assert that inventions worked for the
physical well-being of the employees. Every description of
the drawgirls' task stressed its rigors. The problem was
caused by weights at the base of the loom which made each
tug on the semple very resistant. A fourteen-hour day of
constant labor at the cords wore the young women out in six
or seven years.16 All the inventions to promote the one-
person loom claimed to make the process easy. Each inventor asserted that his technique was "very easy and consequently not at all tiring," "good for the general relief of the
workers," or the means to "preserve the health and strength
of the workers." These assertions were not all formulary.
Contemporaries deplored the physical harm the drawgirls'
jobs inflicted on them. Many expressions of compassion accompanied requests for municipal and state aid for these
women. Nevertheless, there was also an accepted rhetoric
for displaying mechanical virtuosity, and this the inventors
used. Their demonstration of concern for the drawgirls
drew praise. The mechanics were given high marks because
they "deliver our compatriots from these ills by means of
some mechanism easy to move."17

The concern shown for easing physically arduous tasks
was at the positive end of a spectrum of attitudes toward
women workers. The negative aspect was also present, and
inventors used it to discredit the drawgirls and to denigrate
their contributions to cloth manufacture. This was one means
of countering the argument in favor of full employment. Requests for funds to develop the loom sans tireuses characterized the work of the drawgirl not as a necessary expense,
but as "pure loss" (pure perte). It could be expressed this
way only if the function were successfully made part of the
weaver's own regimen. In the petite tire loom which Ponson
invented in 1781, the commissioners of the Lyons Academy
challenged his claim of effectively substituting one weaver's
pedal for the drawing process. "The work is not nearly as
fast as that done when the cords are pulled by hand," they
asserted. "It will not at all give the economy for which [the

inventor] is flattering himself." Until Ponson perfected his
machine, the commissioners implied, the drawgirls' work
and their costs were indispensable.[18]   But only, it was
hoped, until the right technique for eliminating the drawgirl
was found.

   While the emphasis on "wasted money" figured large
in petitions for government subsidy, it also served as a tac-
tic to convince the weavers to invest in the new mechanisms.
If one person could do the work formerly carried on by two
or more, the savings would quickly compensate the master
for the expense of the new device.   They reasoned, unrealis-
tically, that the price of silk stuff, kept artificially high by
the wages of drawgirls, could finally compete with English
cottons and French India prints.

   With so much blame on the expense of drawgirls, it
is necessary to ask how much these auxiliary workers actually cost their employers.   Their critics generally characterized this financial burden in terms of their wages.   But
from other sources we learn that "master-workers ... reduce the salaries of those they employ to the lowest possible
level.   Well, they give only 36 or 40 livres in the form of
wages per year, bread when they care to, a little wine and
three quarters of a pound more or less of meat each day."[19]
We learn that it is not the wage itself that constitutes the
major part of the employers' support, but the cost of room
and board, which runs to one livre a day for women
workers.[20]   To show how much his loom would save, one
inventor totaled up the drawgirl's expenses with this result:
"the yearly wage, the food, the laundry for her bedclothes,
the place to lodge her and the utensils she needs well exceed 300 livres per year."[21]

   Expensive as this was, the Lyonnais understood that
it would have been still more costly if room and board were
not part of the wage.   The syndics and master guards reported with something like horror that drawgirls were in such
high demand in England that they could command wages by a
daily rate, and high ones too.[22]   Yet they still complained
about the high cost of the French drawgirl.   The reason is
found partly in the terms of the girls' employment.   They
were hired by the year and the master workers were obli-
gated to feed them whether or not orders were coming in.
For this reason Philippe De Lasalle called the hiring of
auxiliary workers "a gamble" whose outcome would not be
known for a year.

The master-weavers' financial straits put such expenditures into an unfavorable light and dissipated potential resistance to the new machines. Weavers had always required "the greatest parsimony to exist." During the eighteenth century their condition worsened as their work and wages came increasingly under the large merchants' control. They fought a losing battle to preserve their former influence in the silk manufacturers' guild. They failed to maintain the old guild rules that let masters set their own wages. The potential savings of the loom without tireuses represented one of the few paths by which to get the better of a bad situation. A paradoxical state developed as guild masters invested their hopes in a labor-saving device in order to maintain their economic role as heads of guild workshops.

This role included a virtual monopoly of the weaving trade for the male weavers. Guild rules and inspection kept women from weaving legally unless they were wives, daughters or widows of masters. [23] Although they vaunted the "ease" which their devices brought to the loom, eighteenth-century inventors never suggested that weaving would be easy enough for women to undertake. They wanted to be at the top of their profession, not to be supplanted by the former drawgirls. For guild masters, the loom without the drawgirl seemed to be the means to survival.

However, economic pressures on the French silk industry were becoming too onerous to be overcome by individual initiatives. By the 1780s the quest to replace the drawgirl had become economically obsolete for two reasons: first, because of change in fashion many fewer brocades were being ordered; second, this drastic switch away from the fancy brocades had thrown many drawgirls out of work. They could easily be hired for small wages. The syndics and master guards advised against funding a new loom in 1784 because the device to eliminate the tireuses was no longer economically important. They wrote, "The number of brocade looms is so small today that the suppression of the girls does not present a great advantage for the worker, who can hire them in abundance from Savoy and the surrounding mountains." These officials thought that the English had taken over so much of the brocade industry that they were the ones really in need of the invention. "The manufacture of Spitalfield at London lacks [drawgirls] completely and pays very dearly for the labor it cannot do without," they wrote. "The loom suppressing the tireuses would be far more advantageous to the foreigners than to us."[24]

The change in fashion to cottons, the competition of rival silk works in Spain, England, and Germany, and the heavy tariffs imposed by these countries suggested that the Lyons monopoly of fancy brocade cloth was over. Yet the search for the loom without drawgirls went on. Why? As Abbé Jacquetz put it in his tract on reviving the Fabrique, the brocades had been the most profitable export, far surpassing the plain silk that hardly repaid its expenses. Money from the brocades had underwritten a host of auxiliary trades and provided employment for braid makers, lace makers, designers and those who made other ornaments.[25] Thus businessmen tried to revive the market for this elegant cloth, pressing courts to abandon mourning and petitioning Marie Antoinette to dress in figured silk. The inventors continued to improve the loom à la grande tire at the very time that the Lyons chamber of commerce was trying to revive the demand for products of the machine. The paradigm that machines are created to respond to market needs was reversed in this case.

With business profits declining, those masters with ingenuity sought funds through the avenue of government subsidy. Improvements in loom structure were continually being made in cloth workshops; to formulate a request for official support the weavers had only to describe their new process and argue for its importance. To be sure, they had to satisfy guild officials who judged whether their process was new and whether it actually worked. However, the novelty itself could be an "improvement" rather than a thorough revolution of an industrial process.

In the effort to justify their fiscal requests, the inventors complained increasingly about their female workers. In part they were probably reflecting the views of the weavers who blamed their own workers for general economic hardships that were overwhelming the industry. The new stridency of their comments served to make the self-acting loom seem imperative and to blur the hard-headed view that its use in the silk industry appeared unnecessary. In the light of these motives, the subsidy requests of the 1780s take on a new cast.

Claude Rivey's petition in 1781 set forth a severe indictment of the girls. Rivey's loom was designed to replace the usual complement of thirty-two pedals with a single one and to substitute a simplified cordage that the weaver could activate for the semple. He castigated the drawgirls'

"ineptitude, their weak constitution, their caprices, their illnesses, their stubborn character often opposed to that of an
active and robust male weaver."[26] The girls were head-
strong and insubordinate, to the point of becoming debauched
and taking up prostitution. They fomented "an epidemic of
evil habits in every sense of the word, and often [escaped]
the vigilance of the magistrate [who was] in default, ineffec-
tive, or too indulgent."[27]

The "epidemic" even spread beyond the sphere of
morality as the drawgirls and their cordage were blamed for
crowding up the small apartments, provoking fires, and
spreading mass sickness among the male workers. The
girls themselves were blamed for contracting illnesses that
made them unfit for work and caused them to resort to beg-
ging. Eventually they became charges on the hospitals and
the state. One is reminded here of Barbara Ehrenreich's
and Dierdre English's insight that women are either sick or
sickening, depending on whether they are in the upper or
lower class.[28]

While the women who mounted the design on the loom
were considered unreliable because they came one day and
stayed away the next two, the drawgirl was kept from giving
a full measure of work by the primitive loom technology.
The request for Jacquard's patent license reads like Andrew
Ure's lament that factory work suffers because the operatives
have at least forty seconds between each action.[29] A statement attributed to Jacquard asserted that "in effect the drawgirl only moves after the orders of the principal worker.
These orders are given each time as soon as the shuttle
passes from one hand to the other; but as it is impossible
not to have a certain pause between the order given to draw
this or that buckle and the execution, we must understand
that there is a lot of waste time. These losses should be
measured properly."[30]

IV

When the disruption of the Revolutionary years had
passed and interest focused on rebuilding the silk industry
at Lyons, the loom suppressing the drawgirl once again took
on prominence. What better means of publicizing the silk
industry could there be than issuing a call for this long-sought
invention? In 1807 the Society for the Encouragement of National Industry offered a prize of 3,000 francs for a loom able

to make all types of brocades and figured cloth without the aid of the drawing mechanism. [31]   A preface to the description of Jacquard's new loom reflected the preoccupations of earlier inventors. Along with the old looms' imperfections, the excessive space that the drawgirls occupied was singled out for criticism along with their inefficient work procedures and the obligation to house them.   The old populationist dilemma was even mentioned. [32]

With Joseph Charles Jacquard's invention the loom sans tireuses became an industrial reality.   Jacquard replaced the cylinder with a sliding carriage that rotated a series of pierced cards.   Each card represented the pattern for one pick of the shuttle.   The cards pressed against rows of needles that were attached to hooks governing the warp threads.   The needles that entered holes in the card remained taut and caused the warp threads attached to them to rise.   Needles that did not find holes were displaced and allowed the warp threads attached to them to remain low.   In this way the pattern was woven automatically, since the weaver himself could activate the Jacquard apparatus as well as making the ground cloth. [33]

Jacquard's original loom was quickly improved by a number of skillful assistants.   Though hand looms persisted in many workshops, the Jacquard machines spread rapidly through the weaving centers of France, England, Germany and Switzerland. [34]   The weavers could realize their dream of independent work on a loom that made the creation of patterned cloth a simpler and smoother process.

As looms gradually became mechanized, the weavers no longer set the pace of work for drawgirls, but were themselves harried by an exigent master, machine-paced work. The economic and psychological distress the chefs d'ateliers experienced in the nineteenth century was expressed in labor protests, riots and social criticism.   Bias against women workers persisted.   With the suppression of guilds in 1791 and the eventual spread of looms outside the center of Lyons, many women who might earlier have been drawgirls became weavers.   No longer were drawgirls faulted for caprices, errors, sickness and expense.   These criticisms were leveled instead at the expanded group of women weavers whom the chefs d'ateliers viewed, along with male wage earners, as rivals for employment and workers deficient in professional formation and pride. [35]

Notes

1. This material is discussed in Edwin T. Layton, Jr., "Technology as Knowledge," Technology and Culture, XV (1974), 31-33, 37-38; and see the analysis of Eugene S. Ferguson, "On the Origin and Development of American Mechanical 'Knowhow,'" Mid Continent American Studies Journal, III (1962), 3-16. Other approaches may be found in Melvin Kranzberg and Carroll Pursell, Technology in Western Civilization (New York: Oxford University Press, 1967), I, Introduction. See also the Summer, 1966 issue of Technology and Culture, VII, which treats the theme "Toward a Philosophy of Technology." Also, cf. David Landes, The Unbound Prometheus (Cambridge: Cambridge University Press, 1969).

2. This procedure is discussed in Daryl M. Hafter, "Philippe de Lasalle, from Mise-en-carte to Industrial Design," Winterthur Portfolio, XII (1977), 146-47.

3. Denis Diderot and Jean Le Rond D'Alembert, "Soie," Encyclopédie (Neufchâtel, 1765) XV, 300-301.

4. Maurice Dumas, ed., "Le Tissage et l'apprêt mécanique," in Histoire générale des techniques (Paris: Presses Universitaires de France, 1968), II, 678-679.

5. Ibid., p. 679-680.

6. Alfred Barlow, The History and Principles of Weaving by Hand and by Power (London: Sampson, Low, Marston, Searle and Rivington, 1875), pp. 141, 231, 255.

7. Mercure de France, XLIX (November, 1745), 116-120.

8. Jacques and Nicolas Currat, "Mémoire," Bib. Palais des Arts, Lyons, 110, Fol. 13v.

9. Ibid.; for an assessment of the significance of cam construction in loom development, see Verla Birrell, The Textile Arts (New York: Schocken Books, 1973), p. 223.

10. Letter from M. Gay to "V," January 31, 1770, Arch. Mun Lyons, HH 554, n⁰ 17, Cote AA.

11. Walter Endrei, L'Evolution des techniques du filage et
     du tissage au môyen age à la révolution industrielle,
     trans. Joseph Takacs and Jean Pilisi (Paris: Mou-
     ton, 1968), pp. 147-150.

12. Landes, The Unbound Prometheus, pp. 41-42, 81-88,
     139, 159-166.

13. Abbé Jacquetz, "Mémoire de faire refleurir les manu-
     factures de Lyon," 1784. Bib. Palais des Arts,
     Lyons, 110, Fol. 58, p. 52.

14. Maurice Garden, Lyon et les lyonnais au xviiie siècle
     (Paris: Société d'Edition les Belles-Lettres, [1970]),
     pp. 68-80.

15. Currat, "Mémoire," Fol. 13r and v.

16. Justin Godart, L'Ouvrier en Soie (Lyon: Bernoux et
     Cumin, 1899), pp. 72-73, et passim.

17. "Nouvelles de l'intérieur du Royaume," Gazette de
     l'Agriculture, Commerce, Arts et Finances (Decem-
     ber 11, 1776), p. 818.

18. "Rapport des commissionaires de l'Académie chargés
     d'examiner une machine présentée par le S. Ponson,"
     June 5, 1781, Bib. Palais des Beaux Arts, Lyons,
     189, Fol. 72, p. 76v.

19. "Nouvelles," Gazette de l'Agriculture, pp. 817-818.

20. Excerpt from "Procès-verbal Maîtres Gardes de la
     Grande Fabrique," accompanying a letter from M.
     De Flesseles to M. Trudaine, November 14, 1772.
     Arch. Nat. F^{12} 1444B. He cites the price for the
     liseuses (reader of the mise-en-carte) and the
     faiseuses des lats (makers of the buckles). The
     wage for the latter was 10 sous per day. The
     liseuses worked on a piece basis, earning twelve
     livres for reading a cordage of 600 strings. This
     constituted eight days' work.

21. Report to the "Prêvot des Marchands, Echevin Juges
     de la police des Arts et Métiers," May 29, 1781,
     Arch. Mun. Lyons, HH, No. 40, E, Fol. 235.

22. Report of "Les Syndics et Maîtres Gardes, inspecteurs et controleurs Jurées de la manufacture d'étoffes en argent et soie de la ville de Lyon," on a request presented by Claude Rivey to M. de Calonne, Controleur général des finances, Arch. Mun. Lyons, HH, July 16, 1784.

23. The position of women in the guild is outlined by Godart, L'Ouvrier en Soie, pp. 169-177.

24. Report of "Les Syndics et Maîtres Gardes," July 16, 1784, on Claude Rivey.

25. Abbé Jacquetz, "Mémoire," Fol 59.

26. Claude Rivey, "Report to the Prévôts des Marchands, Echevin Juges de la police des Arts et Métiers," May 29, 1781, Arch. Mun. Lyons, HH. Fol. 235, Liasse 40, Cote E, p. 3v.

27. Ibid.

28. Barbara Ehrenreich and Dierdre English, Complaints and Disorders: The Sexual Politics of Sickness (Old Westbury, Conn.: Feminist Press, 1975), pp. 45-47, 54-59, et passim.

29. The Philosophy of Manufactures (London, 1835), pp. 364-370.

30. Request for Brevet d'Invention of Joseph Charles Jacquard, 5 Vendemiaire, an IX, p. 5. National Patent Office, Paris. Camilo Rodon y Font discusses this comment in L'Historique du métier pour la fabrication des étoffes faconnées, trans. Adolphe Hullebroeck (Paris and Liège: Librairie Polytechnique Charles Beranger, 1934), p. 106. I am indebted to Rita Adrosko, Curator, Division of Textiles, Smithsonian Institution, the National Museum of History and Technology, for a cautionary approach to Rodon y Font's interpretation.

31. The award of the prize to Jacquard was announced in the Bulletin de la Société d'Encouragement pour l'industrie nationale, VII, No. LXVIII (June, 1808), p. 189.

32. These rationales had been made earlier in the Request for Brevet d'Invention, p. 4, 17-18.

33. Daumas, Histoire générale, pp. 681-682; Barlow, History, pp. 140-157.

34. Endrei, L'Evolution, p. 150; Barlow, History, pp. 140-147.

35. Robert J. Bezucha, The Lyon Uprising of 1834 (Cambridge, Mass: Harvard University Press, 1974), pp. 22-29, 35-47; Mary Lynn McDougall, "Working-Class Women during the Industrial Revolution," in Becoming Visible: Women in European History, ed. Renate Bridenthal and Claudia Koonz (Boston: Houghton Mifflin, 1977), p. 259; Laura S. Struminigher, "The Artisan Family: Traditions and Transition in Nineteenth-Century Lyon," Journal of Family History, II (1977), 211-222.

2. LADIES AND LOOMS:

The Social Impact of Machine Power in the American Carpet Industry

by Susan Levine

At first glance, the importance of technological change for women seems to lie simply in particular inventions or mechanical improvements in industries employing women or in the home. On closer inspection, however, the relations between new technologies and women appear not as simple cause and effect, but rather involve changes in the social context of work and new social relations in industry generally. Women and technology both become part of a larger process of social transformation. Neither the entry of women nor the use of new technology alone can account for that transformation, but together they have heralded new eras in American society.[1]

The American carpet industry provides an important example of the process of technological change as it relates to women. As the only branch of the textile industry in the United States with a skilled hand-craft tradition lasting through the late nineteenth century, the carpet industry witnessed not only a transformation in technology with the introduction of power, but a transformation in the appearance and organization of the work force as well. The significance of power in this industry lay not only in the productive superiority of a power loom over a hand loom, but in the relation between the introduction of power and the use of female labor to displace skilled male hand loom weavers.[2]

Carpet manufacture in the United States, unlike the cotton industry, remained dominated by hand production until well after the Civil War. Small manufacturers, skilled hand

loom weavers and small shops characterized the industry.
While carpet manufacture traditionally centered in Philadelphia, as early as the 1850s a few mills grew up in New
York state and Massachusetts. Called "eastern" mills,
these factories were generally larger in scale than Philadelphia's shops and introduced power looms at an earlier
date. However, as late as 1880 hand looms still constituted
one half of all carpet looms operating in the United States.
Despite the size and output of the larger eastern mills, "almost one half the workers and total wages paid, nearly nine-
tenths of the establishments, about one half the value added
by manufacture, [and] three-fifths the total yardage" issued
from Philadelphia until the turn of the twentieth century.[3]

Falling prices after the Civil War, accompanied by
an increased demand for carpets, led manufacturers to seek
ways of reducing costs in production. Power looms appeared
as a means of increasing productivity and at the same time
reducing costs. The new looms could be operated by unskilled or semi-skilled women who demanded a much lower
wage than the skilled hand loom weavers. The introduction
of power, therefore, signaled not merely reduced costs of
production, notably in the cost of labor, but also promised
increased control by the manufacturers over the process of
production which had formerly been the domain of the skilled
weavers. Ultimately, the use of power and the consequent
replacing of skilled weavers by unskilled women resulted
from an interplay of these forces of both economy and social
relations.[4]

Large Civil War profits and a favorable post-war
tariff resulted in a vast expansion of the carpet industry
during the post-war decades. The overall value produced
increased by twenty million dollars between 1860 and 1880,
while the capital invested jumped from five million dollars
in 1860 to twenty-one million twenty years later.[5] Until
the war, domestic production mainly provided a specialty
market while the majority of carpets used in this country
were imported. However, by the mid-1870s domestic manufacturers began to expand their share of the home market
and exceed the level of imports. The industry continued to
expand, and between 1870 and 1890 the capital invested increased three times while the value added went up by one
hundred and twenty per cent.[6]

A dramatic fall in prices accompanied the increased
production. Between 1870 and 1890, prices in the industry,

as in the economy at large, generally fell.   The period of
lowest prices, 1876 to 1879, coincided with the wholesale
introduction of power machinery.   During those years prices
fell by twelve and one half per cent, from highs of $1.70
per yard for medium-grade goods to a low of eighty-seven
cents in 1879.[7]   As prices fell, the demand for medium-
and low-grade goods increased.   Consequently, the record-
low prices of the late 1870s, coming in the wake of technological transition and increased productivity in the industry,
created a glutted market and a further lowering of the price
of goods.   Many manufacturers saw the use of power and unskilled women as the only way to keep costs down and coun-
teract the low prices.[8]

Productivity

        The invention of power looms for carpet manufacture
resulted in a vastly increased output in the industry.   Between 1850 and 1870 power looms were perfected for every
type of carpet from the luxurious Axminster to the popular
ingrain.[9]   These looms as much as tripled output and reduced production costs by fifty per cent.   Where ingrain
hand looms averaged an output of fourteen yards per day,
the new ingrain power looms averaged thirty.[10]   In addition,
the new looms could be run by girls or women at greatly
reduced wage scales.   In 1880 the United States Census of
Manufacturers estimated that "the change from hand looms
worked by men to power looms operated by women has resulted in an increase of one hundred per cent in the productivity and a decrease of over fifty per cent in the cost of
labor."[11]   Alexander Smith, owner of a valuable power loom
patent for Axminster carpets and proprietor of the nation's
largest carpet mill, boasted that "one competent girl will
produce in a day an amount equal to the production of ten
English or French looms attended by as many men."[12]
During the years of lowest prices, from 1876 to 1879, the
number of power looms in Philadelphia alone jumped from
592 to 1,346, and output increased from 130 yards per week
per loom to close to 200.[13]

Wages

        Increased production and falling prices stimulated
economies of production particularly directed at the cost of
labor.   In an industry such as carpet manufacture, where

production was carried out primarily in small, highly competitive shops, the margin of profit was very low. Thus
manufacturers calculated labor as one of their major costs.
With power looms and female operatives, production costs
could be reduced by increasing the speed of output but also,
and even more important in a period of falling prices, by
simply reducing wage rates. Between 1869 and 1879,
weavers' earnings fell by ten per cent. Where hand loom
weavers received up to twenty-five cents per yard, female
power loom operatives averaged five or six cents. With the
increased number of women in the industry and the growing
competition of power looms, even the wages of hand loom
weavers began to fall.[14] Elias Higgins, one of New York's
leading manufacturers, put it simply, "[W]hen you have a
loom weaving thirty yards a day and requiring the attendance
of but one girl ... it is getting the cost of weaving down to
about its lowest level."[15]

Weavers and Their Organization

Philadelphia's hand loom weavers, largely English and
Scottish immigrants, formed a tightly knit community of work
as well as social life in Kensington, Philadelphia's textile
district. During the hand loom era women, often the daughters or wives of the weavers, worked in the trade only as
helpers. Small shops of ten looms or less characterized
carpet manufacture. Skilled weavers, who frequently owned
their own looms, purchased yarn from a carpet merchant
and delivered the finished product to a central factory. Even
after the Civil War, when fewer weavers owned their looms
and increasing numbers went to work directly for manufacturers, the weavers maintained a considerable degree of
autonomy in their work.[16]

The skilled hand loom weaver performed the entire
process of carpet production from "receiving the warp,
beaming it, tying it in, weaving the carpet and taking it out
of the loom, darning and trimming it and assisting in drawing it over and, in many cases, keeping the loom and its
appurtenances in repair."[17] In addition, weavers traditionally set the price for weaving, often including some standard
for materials used as well.

Until the widespread introduction of power looms in
the late 1870s, the skilled craftsmen had been influential in
the trade, particularly with regard to setting wage rates.

As early as 1846, carpet weavers organized in a union which included hand loom weavers in Philadelphia as well as in the eastern mills. Their organization included a constitution for "regulation of the carpet trade" and called for initiation fees for persons wishing to become operatives. Fines for instructing those who did not pay the fee were also included.[18]

This traditional and jealously guarded control of entry into the trade was a major factor in the absence of women from the skilled craft. Although the common explanation for the absence of women was the "heavy weight of the carpet hand loom ... which greatly taxed the arms of the operator," cultural tradition and protection of skill were more likely explanations.[19] A Census observer noted that carpet weavers, "like iron workers and other skilled craftsmen formed a class upon whom the manufacturers were much more dependent than they are at present."[20] And two twentieth-century historians characterized their early union as "an instrument which indicates a spirit of union which is not met with in other branches of the textile industry and is to be found rarely in any American manufacture of that era."[21]

Manufacturers recognized this strength of the weavers, particularly with regard to wage rates or the cost of labor. One manufacturer admitted that "the primary object sought to be gained (in the introduction of power looms) was the better control of weavers and the prevention of strikes."[22] Another commented, "I believe that if we had no strikes there would not have been a dozen power looms in Kensington."[23]

Although weavers maintained their organizations into the 1880s, including two early locals in the Knights of Labor, they could not prevent the introduction of power looms and unskilled women into the trade. Some manufacturers side-stepped the skilled weavers entirely by opening new power factories outside Philadelphia's hand-craft center. The new mills drew from a labor force of young immigrant women in New York and New England towns such as Yonkers, Amsterdam and Lowell.

In the new power mills, the weaving processes were divided into several separate tasks, one operative performing only her assigned operation. Semi-skilled women requiring only a short period of training now worked as loom tenders, setters, winders, spoolers or pickers. In some mills each operation could be separated by as much as a quarter of a mile from the other departments.[24]

However, the new operatives were not totally isolated from the traditional skills of carpet weaving. Loom fixers, the skilled elite of the power factory, retained much of the artisan's overall knowledge of the trade, providing a link between the old work force and the new. Loom fixers set patterns, fixed the warp in the loom and repaired any broken loom parts. The power loom weavers became dependent upon the loom fixers to start and stop their machines and to correct any malfunctions in their looms, warps or patterns. In addition, loom fixers, responsible for twelve or more weavers at a time, often hired the young women themselves. [25]

As power began to dominate the industry, the skill of the hand loom weaver declined in influence. By the late 1870s hand looms produced mainly lower grades of goods or specialty items. Craftsmen's wages fell from a high of twenty-five cents per yard to as little as seven to ten cents. By the 1880s hand loom wages approximated those of the female power loom operatives, who averaged five or six cents per yard. [26] One old weaver told a reporter, "time was when a guid weaver cud make a substantial livin with a hand loom. A body cud raise un's family cunfertable--cud send the children to skule, pay his rent regular and live content. But now that's all changed ... we that's left are jest peggin away like ... at this time o life the likes of us can't learn power loom worruk."[sic][27]

Despite the weaver's reduced stature in the trade, their skill was still essential to the industry. As late as 1886, hand looms still accounted for one-quarter of all looms in operation. [28] During the mid 1880s hand loom weavers in the Knights of Labor engaged in a successful strike demanding an increase in the weaving rate. The power of their skill, combined with effective organization, still proved a difficult adversary for manufacturers who sought control over costs. [29]

Female power loom weavers, of course, could not command the same control over skill as their predecessors in the craft. Learners could be trained to run power looms in as little as two weeks, and operatives could be replaced at the will of the manufacturers. In addition, paternalistic employers and social traditions dictating feminine submissiveness made it difficult for the female operatives to maintain long-term organizations. However, within the limits of technology and social tradition, the women assimilated much of the older craft influences. When the Philadelphia manufacturers, faced by record-low prices, instituted the first wage cut for power loom weavers in 1878, all of the operatives in Kensington went on strike. This action, the first of its kind in the carpet industry, closed many of the city's mills and brought power production to a halt for three months. [30]

The women created a city-wide organization loosely connected to the Knights of Labor. They elected representatives and committees in each shop and paraded in the streets and in front of the mills, pressing their demands. Skilled weavers and loom fixers supported the women in their strike, providing a critical leverage against the employers and hinting at a continuing sense of community in the carpet trade. The women by themselves were painfully vulnerable to replacement by new "learners." The strike ultimately failed because the manufacturers were able to train quickly large numbers of new women. Many manufacturers expected that this advantage "would have a tendency to the repression of future strikes." By the end of the strike, the women estimated that close to one hundred of their members had been permanently replaced. [31]

Despite the failure of immediate demands, this first strike of female operatives in the carpet industry laid the groundwork for future organization in the trade. Manufacturers and weavers agreed to negotiate weaving wages every six months, an unusual step in an industry dominated by unskilled women. By 1883, female carpet weavers formed some of the largest locals in the Knights of Labor and by 1885-86 these locals provided critical leadership for working women during the "Great Upheaval." [32]

Drawing upon many of the traditional prerogatives of carpet weavers, the female operatives began to challenge manufacturers. A long strike of carpet weavers in 1885, beginning in Philadelphia and spreading to all the carpet centers of New York and New England, raised demands not simply for higher wages but for rights and privileges on the shop floor as well. These strikes successfully asserted the women's authority over shop rules, fines and discipline. In some cases the Knights of Labor even attempted to restrict entry to the trade, much as the hand loom weavers had done by refusing to instruct learners who would not pledge to join the Order. [33]

Conclusion

Manufacturers, through the use of new technology, generally succeeded in breaking artisan control in the carpet industry and in limiting the options of the new female workforce. Nevertheless, the women, drawing upon that older tradition both through direct support from skilled males and through indirect transmission of a sense of rights on the job, forged an active role for themselves in the mills. The women were thus influenced not simply by the dominant logic of efficiency associated with establishment of a new technology but also by the countervailing logic of the previous generation of workers in the industry. Created as a result of the new technology, the new generation of workers simultaneously witnessed a breakdown in both the powerful artisan craft tradition and the traditional expectations of women as passive, home-centered beings. Technology proved the catalyst but not the final arbiter of this process.

Notes

1. For a discussion of technology and changes in the nature of work, see Harry Braverman, Labor and Monopoly Capital (New York, 1974); also the debate on worker control, David Montgomery, "Worker's Control of Machine Production in the Nineteenth Century," Labor History (Fall, 1976); and Katherine Stone and Stephen A. Marglin in The Review of Radical Political Economics (Summer, 1975).

2. For a general discussion of the carpet industry, see Arthur H. Cole and Harold T. Williamson, The American Carpet Manufacture: a History and an Analysis (Cambridge, Mass., 1955), and John S. Ewing and Nancy P. Norton, Broadlooms and Businessmen: a History of the Bigelow-Sanford Carpet Company (Cambridge, Mass., 1955).

3. Cole and Williamson, p. 151. Also see J. R. Kendrick, "The Carpet Industry of Philadelphia," Pennsylvania Department of Internal Affairs, Annual Report, Part Three, 1889.

4. See Cole and Williamson, as well as Kendrick. Also see William A. Sullivan, The Industrial Worker in Pennsylvania, 1800-1840 (Harrisburg, Pa., 1955).

5. These figures have been compiled from the United States Census, 1860 to 1890.

6. Victor S. Clark, History of Manufacturers in the United States (New York, 1929), p. 435.

7. The Carpet Trade Review, March, 1880.

8. The Carpet Trade Review, January, 1880. See also March and September, 1880.

9. Cole and Williamson, pp. 57-59.

10. The Carpet Trade Review, January, 1880. Also see August, 1878.

11. Joseph D. Weeks, Report on the Statistics of Wages in Manufacturing Industries, Vol. XX, Tenth Census of the U.S., 1880 (Washington, 1886).

12. John L. Hayes, American Textile Machinery: Its Early History ... (Cambridge, Mass., 1879), p. 53.

13. The Carpet Trade Review, June, 1880.

14. Clark, p. 168. Also see Cole and Williamson, p. 173.

15. The Carpet Trade, July, 1877.

16. Phillip A. Hall, The Rug and Carpet Industry of Philadelphia, (1917), p. 9.

17. The Carpet Trade and Review, September 1, 1883.

18. Cole and Williamson, p. 39.

19. Ibid., p. 25.

20. W. P. Trowbridge, Report on Power and Machinery Employed in Manufactures, Vol. XXII, Tenth Census of the U.S., 1880 (Washington, 1888).

21. Ibid., p. 39.

22. The Carpet Trade, November, 1877.

23. Philadelphia Inquirer, March 5, 1879.

24. Information on the structure of power mills has been compiled from various sources including Cole and Williamson, Ewing and Norton, and contemporary descriptions throughout the Carpet Trade and the Carpet Trade and Review (the two merged in 1882 into the Carpet Trade and Review).

25. Description of loom fixer's job compiled from various sources (ibid.); also, Albert Ainley, Woolen and Worsted Loom Fixing, A Book for Loomfixers ... (1900).

26. Cole and Williamson, p. 173.

27. The Carpet Trade and Review, March 15, 1888. Also see Kendrick, p. 19.

28. Clark, p. 438.

29. See account of strikes in Philadelphia newspapers, especially the Inquirer and the Public Ledger, November, 1885 through June, 1886.

30. See accounts in Philadelphia newspapers, especially the Inquirer and the Public Ledger, November, 1878 through March, 1879.

31. For information on the strike, see Philadelphia newspapers, the Inquirer, Public Ledger, the Press and the Evening Bulletin, November, 1878 to June, 1879. Also see various accounts in the Carpet Trade and the Carpet Trade and Review, same dates.

32. For more information on the strike agreements, see Cole and Williamson, pp. 184-186. For more on women carpet weavers in the Knights of Labor, see Susan Levine, "'Then Strive for Your Rights, O, Sisters Dear,' The Carpet Weavers' Strike and the Knights of Labor," unpublished, mimeo, 1976.

33. Cole and Williamson, pp. 184-186; also Levine, ibid. For a general discussion of the issues of fines and discipline, see Susan Levine, "Honor Each Noble Maid, the Carpet Weavers' Strike in Yonkers," unpublished, mimeo, 1976.

3.   TECHNOLOGICAL CHANGE AND WOMEN'S WORK:

Mechanization in the Berkshire Paper Industry,
1820-1855*

by Judith A. McGaw

I

Most of the conventional wisdom about American
women's industrial work in the nineteenth century rests upon
the supposition voiced by Caroline Ware nearly fifty years
ago that:

> The story of the New England cotton industry
> is the story of the industrialization of America.
> This industry brought the factory system to the
> United States and furnished the laboratory wherein
> were worked out industrial methods characteristic
> of the nation.[1]

Depending heavily upon studies of the cotton textile industry,
we have assumed that the advent of machine production dramatically altered women's economic roles.

We can summarize the consensus. At first, technological change drew women into factories to tend machines
performing tasks previously carried on around the family
hearth or to replace craftsmen as unskilled or semi-skilled
machine tenders. These women were young, unmarried,
middle-class, and worked only for a short time. After an
initial period of more or less enlightened paternalism, conditions of work deteriorated, wages fell, and machines ran

*An earlier version of this paper was presented at the De-
cember, 1977 meeting of the American Historical Association. It has benefited from readings and comments by
Brooke Hindle, Robert Asher, Maurine Greenwald, Ronald
Snell, and Martha Moore Trescott.

more rapidly. Consequently, the middle class no longer
countenanced factory labor and its daughters relinquished
their places to their immigrant, working class sisters.
Like early working women, the lower-class mill girls of
the later nineteenth century were young and single. They
only worked for a few years to help their families before
marrying. Since few scholars have studied the machine's
impact on women's work in industries other than textiles,
only an occasional demurrer has been registered against
this view. [2]

This paper looks at the introduction of machinery in
the nineteenth-century American paper industry and at its
impact upon women and women's work in America's leading
paper-making region during the period. While the case of
paper making supports certain elements of the consensus, it
takes exception to others. Women worked in paper mills
before mechanization. Machines brought virtually no change
in the kind of work women performed or in the division of
labor between the sexes. Women continued to hold unskilled
jobs and to receive less pay, but in other respects their
jobs became more desirable than those held by men. No
pay reductions, deteriorating conditions, or speed-ups drove
single, middle-class, native-born women from the mills.
They remained, while an increasing number of jobs for
women drew many additional workers into the mills, including
Irish and French Canadian women and married women. The
heavy demand for women's labor, among other factors, made
women workers as likely to remain on the job as men. [3]

These conclusions are based upon a study of machines
and workers in the fifty-two paper mills operating in Berkshire County, Massachusetts, between 1820 and 1885. This
region is a useful choice for study, not only because it was
the nation's premier paper-making region during the industry's mechanization, but also because it offers a large body
of preserved records, including photographs of mill labor as
workers performed it throughout most of the nineteenth century. Company records, contemporary accounts, and these
photographs make obvious the limited impact of technological
changes on women's work and on the sexual division of labor
in the paper industry. This can be seen most clearly by
looking at the various steps in paper production. [4]

Figure 1 (above)   Figure 2 (below)

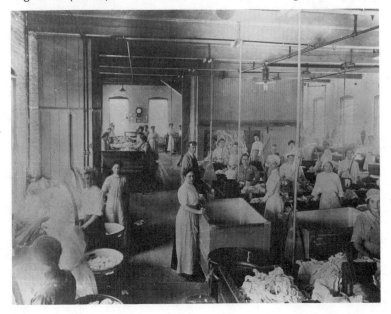

Figure 3 (above)   Figure 4 (below)

II

Paper making in most nineteenth-century Berkshire mills began with the arrival of rags at the mill. Since unloading, breaking up, and shaking debris out of these huge bales was heavy and dirty work, it, like other tasks of this character, was performed by men.[5] This was true both before and after the introduction of the mechanical rag thrasher shown in the background of Figure 1.[6]

Upon completing this task, the men delivered the rags to the rag room (Figure 2). Here some women sorted rags by color and texture, while other women, using scythes mounted in front of them, cut out defective sections and reduced the rags to small, easily processed pieces. The scarcity of labor and great demand for paper had drawn these women into the mills long before the introduction of machinery. The machines' arrival affected them only by increasing the demand for rags and, consequently, for rag room workers. Throughout the years from 1820 to 1885 the majority of women in the industry worked here, at tasks requiring neither skill nor strength nor initiative. As in other departments of the mill, the supervisor of this all-female department was a man.[7]

From the rag room, men carried the rags to the beater room (Figure 3). Here huge wooden contrivances called rag engines washed the rags and ground them to fiber. Installed in most American mills before the beginning of the nineteenth century, these engines needed a skilled operator since the treatment of fiber here determined the final character of the paper. Throughout the nineteenth century, therefore, the engineer, who presided over this process of beating, was a man. Working virtually alone, he had considerable initiative, another characteristic of men's, not women's, work in the industry.

After beating, the rag fiber was ready to be formed into sheets of paper. The outstanding technological advance in the nineteenth-century paper-making industry occurred at this stage of production. Beginning in 1827 in the Berkshires, the cylinder paper-making machine replaced the skilled vatman who had previously formed individual sheets of paper on a hand mould. Then, starting in 1843 in the Berkshires, the Fourdrinier paper-making machine, such as the one shown in Figure 4, generally superseded the less satisfactory cylinder machine.[8] Far from making it possible,

however, for unskilled female machine tenders to replace
skilled men, the paper-making machine required great skill
from its tenders. Its intricate series of rotating cylinders
and gears and the complex mechanism shaking the forming
paper from side to side needed frequent adjustment to prevent the endless sheet of paper from tearing, backing up on
itself, or running too light or too heavy. Consequently,
like the vatmen they superseded, the machine tender and
back tender were highly skilled males. Characteristically,
their work also demanded physical strength and initiative.

Since Berkshire paper mills installed various sections
of the paper-making machine at different times, gaps in
mechanization created short-lived positions for women. For
example, until the layboy (a device which removed and
stacked the sheets of paper) was perfected in 1847, women
sat at the end of the machine and caught and stacked the
sheets by hand. Elsewhere, until mills began to add steam
dryers to their machines in 1834, women and men hung the
paper to dry over wooden poles in dry lofts (Figure 5). In
some Berkshire mills where particularly fine grades of paper
were made, this practice was continued well into the twentieth century. While the sexes worked side by side in the
dry loft, tasks requiring judgment or supervision, such as
deciding where paper should be hung and when it should be
taken down, and tasks requiring strength, such as carrying
the heavy stacks of paper to and from the loft, fell to men.
Women performed menial unskilled work, such as tapping the
edges of the stacks to even them.

The finishing room employed most of the women not
working in the mill's rag room. Berkshire's fine paper
mills guaranteed their customers perfect paper, so most
women finishers, like those seen on the left in Figure 6,
sat at long tables and scanned the paper, sheet by sheet,
for flaws. As with rag room workers, the advent of machine production affected these women only by increasing
their number. Most of the men in this section of the mill
performed the heavy work of trimming and packing the paper.

The finishing room was, however, the one section of
the mill where the introduction of new machines changed
women's work. As shown in Figure 7, women prepared
stacks of paper and metal or cardboard sheets which stronger
and more skilled men ran through the new platers to give the
paper distinctive surfaces. The calender, another paper-sur-
facing machine, required little strength and only a modicum

Figure 5 (above)   Figure 6 (below)

Figure 7 (above)   Figure 8 (below)

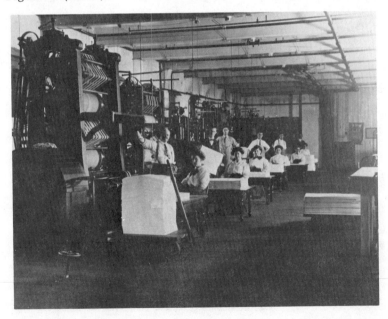

of skill to attend. Therefore, as shown in Figure 8, women sat on either side of these machines, one feeding in sheets of paper and the other removing them. Similarly, pairs of women attended ruling machines. Both calenders and ruling machines created these new positions for women in the Berkshires beginning in the 1830s. Beginning in the 1840s, individual women operated other new finishing room machines which die-stamped paper and converted it into envelopes (Figure 9). These, like other women's machines, did not require skilled attendants. Men performed the skilled operations of adjusting the machines to handle different thicknesses of paper or to rule different patterns of lines. Female machine operators were the exception to the rule, however. Most finishing-room women continued to perform unskilled hand labor, like assembling stationary boxes (Figure 10) and counting and folding paper.

III

Just as it merely confirmed the sexual division of labor along lines of skill, mechanization simply perpetuated earlier substantial differences in remuneration.[9] Although both men's and women's wages rose continuously in this industry throughout the nineteenth century, women, both before and after the introduction of machinery, earned only one-third as much as men on the average. In part, the differential resulted from the contemporary practice of paying women less than men simply because they were women. Female teachers in Berkshire paper mill towns, for example, earned only half as much as males received for the same work. The fact that female paper mill operatives earned proportionately less reflected their less skilled tasks as well.

Not only were they paid less; unlike men, whose wages varied considerably with their varying levels of skill, women workers earned virtually uniform pay for their uniformly unskilled tasks. Thus, while a woman who greatly exceeded the required quantity of work sometimes earned as much as the least skilled man in the mill, she lacked his expectation of rising to a more highly paid position. It is hardly surprising, then, that women saw little difference among the various jobs open to them. Less than half of those who wrote applying for jobs in the mills bothered to request a particular position. By contrast, men almost invariably asked for a specific job and cited their qualifications.[10]

Figure 9 (above)   Figure 10 (below)

In paper making, therefore, the introduction of machines was not responsible for bringing women into the mills or placing them in unskilled and poorly paid jobs. Those decisions had been made before the machine arrived. Rather, because mechanization particularly affected men's work, the deteriorating working conditions that accompanied machine production enhanced women's work by contrast. They came to work shorter hours under safer, more pleasant working conditions. [11]

IV

Before the advent of the machine, paper workers, both male and female, had had some control over their hours of work. Most had worked under a system which defined a certain number of sheets of paper or pounds of rags as a day's work. At most, workers put in an eleven-hour day, arriving at six and departing at six, but taking time out for breakfast, dinner, and grog breaks. After the introduction of the new machinery, women continued to work on this basis and the amount of paper or rags to be processed for a day's work remained the same. Undoubtedly because the region's fine paper mills relied upon their working women's care in sorting rags and paper to maintain the quality of their product, they did not resort to speed-ups. Since most women routinely accomplished one and a half times the required work, they probably could have worked fewer than eleven hours. Moreover, the nature of women's work permitted flexible scheduling to meet individual needs. At least one worker recalled that married women, many of whom probably had children in school, left at four in the afternoon after working only eight hours. While these women were the exception, the years following mechanization brought shorter hours for all women. Mills in at least one Berkshire town shifted to the ten-hour day for women even before state legislation mandated it. Several years later, Saturday work hours were reduced to nine. As their work week fell from sixty-six to fifty-nine hours, these women's real incomes rose.

Men who worked with women in the rag room, finishing room, and dry loft shared the benefit of these shorter hours. The principal male employees, however, tended the expensive new paper machines, the costly steam engines and boilers introduced to supply power and steam, and the increasingly large number of high-priced rag engines added to

grind pulp for the machines. To defray their cost, these
machines had to operate long hours. The 1840s found some
men working as many as fifteen hours a day. Within a few
years, however, mills had installed artificial lighting and
were running day and night. For the rest of the century,
most men worked twelve-hour shifts, alternating between a
night shift one week and a day shift the next.

Women not only had more pleasant work hours, they
spent those hours in more pleasant surroundings. Separated
from the mill's heavy machinery, they worked in comparative quiet. Those women who tended finishing machines
either worked side by side or in pairs, which allowed them
to converse with one another. The vast majority of women,
those in the rag room and those sorting paper, worked in
large groups where they could talk freely to relieve the
monotony of their tasks. This ability to tolerate tedious
work by talking and working at the same time may have
contributed to women's retaining a monopoly on these jobs.
At least one mill superintendent tried young men at paper
sorting, but found that when they began to talk their hands
slowed. Perhaps experience in performing monotonous, but
gregarious, household chores had helped women develop this
ability.

Whether or not this was the case, the increased noise
in the beater rooms and the even greater noise level of the
machine room made it necessary for men to shout to be
heard. This undoubtedly kept conversation at a minimum.
Also, while women's work in the mills was gregarious,
men's was solitary. Figures 3, 4, 11, and 12 depict this
isolation of the male paper worker vividly.

The physical conditions male workers endured were
also less desirable than those accorded women. The odor
of lard oil, the principal lubricant, and the smoke of kerosene and coal oil lanterns fouled the air in the unventilated,
continuously operated machine rooms. Water, evaporating
from the heated rag engines and paper machines, condensed
on the cooler walls and ceilings. One worker seeking employment elsewhere explained, "My object in leaving this
place is on account of the Room in which I work, it is verry
[sic] wet and it is ingering [sic] my health." After quoting
his wages he concluded, "so you see my wages is good--but
money will not purches [sic] health." Women undoubtedly
ran a risk of respiratory disease from working in the dusty
rag rooms, and in at least two cases they contracted smallpox

Figure 11 (above)   Figure 12 (below)

from the rags of an infected person. However, in general, their working conditions were preferable to those of their male counterparts. [12]

Not only did mechanization subject men to less healthy conditions; it also meant that they faced the risk of bodily injury, something virtually unknown in the hand-made paper era. In this respect, as in others, women's working conditions remained unchanged by mechanization and were, therefore, much safer. According to reports of mill accidents between 1857 and 1871 in the newspaper of Berkshire's leading paper-mill town, nearly seven times as many men were injured on the job. Men also suffered seven times the number of fatal injuries. Many men had fingers and hands cut and crushed in the new machinery. Male workers who were engaged in construction or repairs around the mills frequently fell and suffered bruises and broken bones. Almost as frequent and much more terrifying were the many accidents caused by the flying belts and whirring shafts near which men worked (Figure 13). Over half of the fatal accidents befalling men occurred when they inadvertently came in contact with these. Indeed, of the few accidents involving women during these years, nearly half took place when they ventured out of their work rooms into men's areas and came too close to revolving shafts. Steam boiler explosions, another hazard female workers were not exposed to, also killed or maimed several male workers during these years. [13]

V

Both because machine production did not adversely affect women's work and because it increased the demand for their labor, no mass exodus of native-born, middle-class women from the mills occurred in the Berkshire County paper industry. In 1865, after immigrants had replaced the Lowell girls, a Berkshire mill owner had two applicants from a neighboring hill town recommended to him as "good respectable Yankee girls." Five years later nearly half of the women working in the six paper mills of Dalton were native-born daughters of native-born parents and half of the rest were native-born of foreign-born stock. In the three mills in Adams to the north, more than half of the women were the native-born children of native-born. The mills of Lee and New Marlborough to the south employed more women who were foreign-born or the offspring of foreign-born, but even here about one-fourth of the female work force was native-born of native stock. Since the female work force had grown substantially in the years following mechanization, the absolute numbers of native-born workers may well have remained the same or increased. The percentages of foreign-born male workers were substantially higher in all four towns, suggesting that where machine production had adversely affected work, native workers may have left the mills.[14]

The fact that women's work was relatively pleasant and greatly in demand also meant that the shift in American middle-class values toward disapproval of women's work outside the home operated less forcefully in Berkshire County. While middle-class women may have ceased working in mills elsewhere, this is true in Berkshire County only if all wage earners are consigned to the lower class. Within the context of Berkshire mill-town society where, except for a few mill owners, storekeepers, and independent craftsmen, all residents were either small farmers or mill hands, highly skilled, highly paid paper machine tenders and engineers ranked as members of the middle class. They were respected citizens of the community and took great pride in their work. The fact that paper mill owners had to master these workers' skills in order to succeed in their businesses and the fact that machine tenders and engineers socialized with and sometimes became close personal friends of mill owners also contributed to these positions being perceived as middle class. Yet the wives and daughters of machine tenders and engineers worked in the mills well into the 1880s.[15]

Individual cases also indicate that mill work remained respectable for middle-class women. More than ten years after paper machines were installed, Eliza B. Thompson worked in one Berkshire mill until the month before she married its owner. As late as 1883 Ellen M. Tufts wrote from Boston to a Berkshire mill owner: "I noticed that you employ ladies in your business," she commented, "have you any vacancies at present or are you liable to have any? I am a graduate of Mt. Holyoke Seminary but being a little deaf I cannot teach school and am at present supporting myself by my needle for which work the pay is small." She concluded by giving the Rev. Phillips Brooks as a reference.[16]

The increased demand for women workers which accompanied machine production probably also accounted for the increasing proportion of married women workers, another respect in which Berkshire paper workers differed from their sisters in the textile mills to the east. One company's mill books show less than five per cent of the female work force married in the early 1820s, but nearly half of it married by the 1870s. Differences between towns in the percentages of women workers who were married suggest that an inadequate supply of single women accounted for married women entering the mills. In New Marlborough, a small isolated town with a stable or declining population and an increasing number of mills and mill workers, single women paper workers were in the minority by 1870. Thirty-nine per cent of the working women were currently married and eighteen per cent were widowed or deserted. In Dalton--which could draw upon an increasing population, the adjoining large town of Pittsfield, and the neighboring hill towns for workers--only four per cent of the women were not single. Lee and Adams, with fifteen and twenty-three per cent of their working women married or widowed, fell in between. [17]

A similar pattern is evident in the age distribution of women workers in different county paper-mill towns. As in the textile mills to the east, a majority of working women in all of the towns fell between the ages of fiteeen and thirty. However, in New Marlborough, with its limited population resources, nearly one-quarter of the women paper workers were forty years of age or older. Mills in Dalton, on the other hand, drew only five per cent of their population from this older group. Again, Lee and Adams, with eleven and fifteen per cent of their working women aged forty and older, fell in between. [18]

The faces of workers in a single Dalton mill, shown in Figures 13 through 16, tell the same story, but more vividly. While skilled men held a number of positions, the majority of the workers were women. And, although many of these women were young, a large number could not properly have been called mill girls.

The increased call for women workers that accompanied mechanization in the Berkshire paper industry and the relative pleasantness of the work undoubtedly account for a final way in which females employed in these mills departed from the stereotype of the nineteenth-century woman worker. Rather than being a markedly less permanent member of the work force than her male fellow worker, the typical single female was only slightly less permanent a member of the labor force than the average man. The average married woman worker, despite more frequent interruption of her career, spent more total months in the mills than either men or single women. [19]

Figure 13 (above)   Figure 14 (below)

Figure 15 (above)   Figure 16 (below)

All workers averaged a shorter tenure in the mills after the advent of the machine, perhaps simply because the increased demand for labor drew more marginal workers into the mills. Men, who worked most closely with the new machines, suffered the most precipitous decline in persistence. The staying power of single women, who more frequently attended the new finishing room machinery, declined less markedly. Married women, those least likely to work with machines, showed the least change in persistence and, consequently, became the most stable element in the work force in the years after mechanization.[20]

VI

This look at women workers in the nineteenth-century American paper industry calls into question the consensus portrait of the typical female factory operative and the impact of machine production upon her. The case of the paper industry makes clear that light, repetitive tasks requiring neither skill nor initiative had been defined as women's work before the advent of machine production. Most paper machines required skilled attendants, so their introduction had little direct impact on women's work.

Indirectly, women benefited from the machine's use. Machines increased the demand for female labor enough that wages rose for the rest of the century. The demand for women workers also drew increasing numbers of married women, older women, and immigrants into the mills. On the other hand, because women, unlike men, worked in relatively safe, pleasant conditions for shorter hours, middle-class, native-born women continued to work beside these new arrivals. Consequently, although many women paper workers resembled the textile worker and the consensus mill girl created in her image, judging from Berkshire County, Massachusetts, the female paper-mill work force was far more diverse. Additional studies may well reveal that the nature of the woman mill worker and the impact of machinery on her work varied substantially from industry to industry.

Notes

1. Caroline Ware, The Early New England Cotton Manufacturers: A Study of Industrial Beginnings (Boston, 1931), 3.

2. Elizabeth Faulkner Baker, Technology and Woman's

Work (New York, 1964) is the best recent summary
of the impact of mechanization on female industrial
workers. Other than textile workers, most of the
women in manufacturing surveyed worked not in
factories, but at home in sweated industries or
through the putting out system. Baker's limitations
merely reflect the limitations in the monograph
literature. Changing attitudes toward women's
work are summarized in Gerda Lerner, "The Lady
and the Mill Girl: Changes in the Status of Women
in the Age of Jackson," in Jean E. Friedman and
William G. Shade, Our American Sisters: Women
in American Life and Thought (Boston, 1976), 120-32.

For more recent discussions, see also
Thomas Dublin, "Women, Work, and the Family:
Female Operatives in the Lowell Mills, 1830-1860,"
Feminist Studies, 3 (Fall, 1975), 30-39, and "Women,
Work, and Protest in the Early Lowell Mills," Labor
History, 16 (Winter, 1975), 99-116; also, Daniel J.
Walkowitz, "Working Class Women in the Gilded
Age: Factory, Community, and Family Life among
Cohoes, New York Cotton Workers," in Friedman
and Shade, Our American Sisters, 179-202.

Studying early twentieth-century textile workers,
Tamara K. Hareven, "Family Time and Industrial
Time: Family and Work in a Planned Corporation
Town, 1900-1924," Journal of Urban History, 1
(May, 1975), 365-389, suggests that by the later
period single women workers may have been the
most permanent mill employees.

For some indication of the limited recent research on the characteristics and work experience
of female industrial workers, see Barbara Sicher-
man's review essay, "American History," in Signs,
1 (Winter, 1975), 461-485, and Dorothy Swanson,
"Annual Bibliography on American Labor History,
1975," Labor History, 17 (Fall, 1976), 586-605.

3. In addition to sources cited below in the fuller discussion of these generalizations, this study has relied
   upon extensive paper company records, especially
   time books, company story daybooks, interviews with
   old workers conducted in the 1930s, and letters
   written by workers applying for jobs. For a complete list of sources employed and a description of
   the region and mills studied see the author's The
   Sources and Impact of Mechanization: The Berkshire
   County, Massachusetts Paper Industry, 1801-1885
   as a Case Study (unpublished Ph. D. dissertation, New
   York University, 1977), 1-7, 386-94, 401-24, 437-47.

4. The photographs presented with this paper have been se-
    lected from a much larger collection at the Crane
    Museum of Paper Making, Dalton, Massachusetts.
    About half were taken in 1892 and about half in 1913.
    The only later photograph (Figure 5) was taken in
    1939, but represents a hand process carried on as
    it had been since colonial times. Although most of
    the photographs, therefore, depict a period somewhat
    later than that under discussion, a careful study of
    the evolution of paper machinery and the composition
    of the work force has ascertained that the same ma-
    chinery, tasks, and division of labor existed in the
    period under discussion. Only peripheral technology,
    such as the electric lights pictured, is unrepresenta-
    tive. See McGaw, Sources and Impact, 67-125.
    For further evidence of the persistence of the sexual
    division of labor outlined below, see B. L. Hutchins'
    study of British paper mill work in 1904: "The Em-
    ployment of Women in Paper Mills," The Economic
    Journal, 14 (June, 1904), 235-248.

5. Except where indicated this and subsequent descriptions
    of work roles rely upon McGaw, Sources and Impact,
    40-53, 60-61, 306-328, 344-347.

6. Precise dating of the arrival of this and other machines
    in Berkshire mills is difficult since the fifty-two
    county mills varied considerably as to when they
    adopted different machines, depending upon when
    they were built, their location, their capitalization,
    the cost of and expected savings from particular ma-
    chines, and the type of paper a mill specialized in
    making. In general, mechanization in the industry
    began with the introduction of the first paper-making
    machine in 1827 and was quickly extended to pre-
    ceding and succeeding stages of production. Most
    mills had adopted all of the machines mentioned in
    the following discussion by the late 1850s. For a
    detailed discussion of this process see McGaw,
    Sources and Impact, 69-125.

7. A few of the regions' mills introduced the use of me-
    chanical rag cutters or wood pulp fiber toward the
    end of the period. These, of course, eliminated
    women rag cutters' jobs. Since the Berkshire in-
    dustry increasingly specialized in fine paper produc-
    tion, however, most mills rejected these innova-
    tions. Only electronic technology and improved rag

cutters introduced in the twentieth century led to the large scale elimination of women's jobs.

8. In the interests of brevity, this statement greatly over-simplifies the technological changes involved. For a full discussion, see McGaw, Sources and Impact, 67-125.

9. For a full discussion of changing men's and women's wages in the industry, see McGaw, Sources and Impact, 328-344.

10. The letters referred to are those written to and pre-served by two Dalton paper companies. However, the fact that these mills specialized in fine paper making and hired proportionately more semi-skilled finishing room women should, if anything, provide more cases of women requesting such jobs, if they had been seen as preferable.

11. Most of the discussion which follows relies on a de-tailed description of working conditions in McGaw, Sources and Impact, 319-328.

12. Letter in "Employees" file, Crane Museum of Paper Making, 4/3/1849; The Valley Gleaner, IX (August 24, 1865), 2; Memorandum from Frank J. Kelly to author, July, 1977.

13. All issues of The Valley Gleaner, a weekly newspaper published in Lee, Massachusetts, have been read, beginning with Volume I, 1857 and continuing through Volume XV, 1871. All reported paper mill acci-dents occurring in Berkshire County have been re-corded. A total of fifty-four accidents, most of them in Lee mills, was reported: forty-seven in-jured men and seven injured women.
    There is no evidence of a tendency to ignore women's accidents, since even rather minor injuries are recorded. Missing, however, are instances of long-time women paper sorters snapping tendons after years of rotating their wrists through the same motion, a work-related injury among sorters in con-temporary mills. Also missing are paper cuts which must have been ubiquitous among finishing room workers, the majority of whom were women.
    Since female workers outnumbered males throughout the period, the chances of injury to male

workers were more than seven times as great as for females.

14. "Employees" file, Crane Museum of Paper Making, 12/14/1864; statistical data is based upon an analysis of the ethnicity of all individuals listed as paper workers in the 1870 United States Census schedules for Adams, Dalton, Lee, and New Marlborough.

15. McGaw, Sources and Impact, 266-270, 314-316, 349.

16. Zenas Crane Time Book, 1836-1848, Crane & Co. Office, passim; "Employees" file, Crane Museum of Paper Making, 11/26/1883. The fact that Tufts saw paper-mill work as an alternative to school teaching, an acceptable occupation for middle class women, is particularly illuminating.

17. Zenas Crane Time Book, 1809-1836, Crane & Co. Office, passim; Crane & Co. Time Book, 1863-1876, Crane & Co. Office, passim; analysis of women listed as paper workers in the 1870 United States Census schedules for Adams, Dalton, Lee, and New Marlborough. Mill time books and employee recollections both show much higher proportions of married women workers in Dalton than do census schedules. Perhaps male workers believed it more respectable to claim sole support of their families and, consequently, lied to the census takers. Census takers' assumptions about married women may also have led to failure to ask their occupations. Differences between towns would then be explicable by the fact that in smaller towns the census taker was more likely to know the family, making lying or errors less common.

18. Dublin, "Women, Work and Protest," p. 106; analysis of women paper workers in the 1870 United States Census schedules for Adams, Dalton, Lee, and New Marlborough. The possibility of selective reporting noted above would also affect age distributions.

19. For graphs based on company time books and a discussion of these changes see McGaw, Sources and Impact, 364-371.

20. Ibid.

INTRODUCTION TO PART I(B)

Women Inventors, Engineers, Scientists and Entrepreneurs

The first two papers in this section deal with groups of women, as in Part I(A), while the third highlights an individual woman. Deborah J. Warner, of the Smithsonian, has studied the women inventors who exhibited their inventions at the Centennial celebration in 1876. She has found that the hope of profit generally motivated these inventors, who produced a number of pragmatic and, at times, commercially marketed inventions. Margaret W. Rossiter, of the University of California at Berkeley, examines the careers of over five hundred female scientists in the United States prior to 1920 and provides statistical data on fields of science these women entered, the number who held Ph. D. s, and where they were employed, among other topics. She finds that discrimination against women in science operated to a substantial degree; Warner notes discrimination against women inventors also.

Martha M. Trescott has similarly noted discrimination against the contributions of Julia B. Hall, sister of Charles Martin Hall, whose inventions helped found the forerunner of the Aluminum Company of America. Julia was not recognized for her work by her brother, nor has the significance of her efforts been properly discussed and documented by Charles's biographer and other historians. While this essay focuses on an individual, Julia Hall should be seen as prototypical of many other female relatives of inventors, women who assisted the invention and innovation process in various ways as it took place in or near the home.

The work on the Halls has brought to light in Trescott's subsequent work various other women who helped effect technological change, including inventors, scientists and engineers. Margaret Rossiter continues her research on women scientists and engineers in America from the early nineteenth century through 1975. Also, Warner and Rossiter will consult with Trescott on a history of women engineers in the U.S., along with Carroll W. Pursell, Jr., who has researched aspects of this topic. As yet, the history of women engineers is not well known, as Rossiter has implied in her essay.

The women depicted in this section cannot be classed, in Cowan's term, as "anti-technocrats," for they entered into and contributed to the scientific and technical communities of their day. However, because science and technology were (and continue to be) heavily male-dominated, it is fair to say that their contributions have mostly not been well known and the impact of their work not felt as it might have been had females in science and technology been given more credibility.

It is interesting to speculate on the different paths technological change might have taken at various points had the real and potential contributions of these and other women been taken more seriously. Professor Cowan is undoubtedly correct in implying that as women move into more major roles in decision-making about technological change, society will begin to see different products from those men would design, produce and consume. (This is already occurring, for instance, in the area of contraceptive technology, with the help and advice of women gynecologists.)

1. WOMEN INVENTORS
AT THE CENTENNIAL

by Deborah J. Warner

The Women's Pavilion at the Centennial celebration, held in Philadelphia in 1876, signaled the entrance of skilled crafts-women into the world of industrial fairs of nineteenth-century America.[1] There had been many previous fairs, both local and international, and always they had welcomed exhibits by women. But these earlier exhibits, for the most part, represented traditional women's work, and the few industrial objects devised by women were lost amid the multitudes of such things designed and made by men. In Philadelphia in 1876, for the first time, there was a separate building devoted exclusively to the wide-ranging products of women's "thought and labor." Invoking Biblical sanction, the women's motto was "Give her of the work of her hands, and let her own works praise her in the gates." These lines from the Thirty-First Proverb were emblazoned over each entrance to the Pavilion.

Some of the exhibits reflected amateur, domestic or charitable activities, but the great majority exemplified things women did for money. Politics, in terms of the suffrage campaign, was strictly excluded,[2] but the equally political struggle for equal employment opportunity and fair compensation was evident throughout. In the words of Mrs. E. D. Gillespie, Chairman of the Women's Centennial Committee, the exhibit aimed "to give to the mass of women who were laboring by the needle and obtaining only a scanty subsistence, the opportunity to see what women were capable of attaining unto in other and higher branches of industry."[3] New Century for Woman, the newspaper published by women in the Pavilion each week during the Centennial season, was dedicated to "the industrial interests of women," and it helped carry the message far beyond the confines of the fairgrounds in Fairmount Park.

Behind the patriotic fanfare, the Centennial was primarily a trade fair where inventors seeking backers or outright buyers for their inventions, no less than established manufacturers, came to show and sell. These trade fairs, of which the Centennial was simply the most extravagant in the U.S. to that date, were the mass-advertising gimmick of the age, offering the widest exposure to the most interested audience at a nominal cost. The opinions of the judges were probably the most objective evaluations made of these objects, and they were cited in advertisements for years to come.

Of all the "higher branches of industry" open to women, invention was one of the most promising. Mrs. Gillespie let it be known that inventors would be especially welcome as exhibitors, and some seventy-nine women responded to her call, filling one-quarter of the total space in the Women's Pavilion.[4] As in the case of Mary Evard, whose Reliance Cook Stove came to the Centennial with notice of its silver medal won at the Missouri Agricultural Exposition of 1873, at least twenty-two of these women showed their inventions at one fair after another.

The fifteen non-domestic inventions by women included signal flares used by the Navy, a life-saving mattress certified by the Board of the United States Supervising Inspectors of Steamboats, an eccentric gauge for adapting candy machines to cut different forms of lozenges, and a variety of educational aids.

The great majority of the women's inventions, however, were domestic. There were six instruments for washing and ironing clothes, eight utensils for sewing, seven dress-cutting systems, six corsets, 18 pieces of furniture or building components, five implements for flower or hair decoration, two medical devices, and 13 kitchen utensils, including three stoves, two dishwashers, and various hand tools. Some of these seem to be little more than ingenious forerunners of the Vegematic, but many were considered by the Centennial judges to represent positive steps towards alleviating the "tedious labor of this particular branch of housewifery." All told, at least fifteen of the women received Centennial awards for their inventions.

None of the women's inventions was important in any large social or technological sense, but many were successful on their own terms. Whatever its actual validity, the myth that invention was a sure path to financial success, for women as well as men, was widely believed. By 1900 fully 75 per cent of the patents granted to women during the previous five years were reputedly yielding profitable returns.[5] Of the inventions by women shown at the Centennial, at least 35 per cent (28 of the 79) were actually produced. And to judge from advertisements and other exhibition catalogs, several of these were in production for at least twenty years.

As a vocation for women, invention was widely and highly recommended. For more than twenty years the Scientific American, our leading advocate of science as useful knowledge, had been running frequent articles about women inventors and ads from the Munn & Co. Patent Agency encouraging women to bring forth their inventions. In 1873, after noting with approval the appointment of the first female first assistant examiner in the Patent Office, the Scientific American had gone so far as to suggest that the radical feminist, Susan B. Anthony, be appointed Patent Commissioner. A faithful champion of women's participation in science, technology and business, yet always mindful of innate physical differences between the sexes, Scientific American noted that there was nothing inherently unladylike about the process of invention. Like novel writing, it could be done in the parlor at home, and did not require traffic in the factory or marketplace.[6]

The first American patent to a woman had been granted in 1809, and the numbers remained small until after the mid-century passage of the various laws which gave married women control over their own property and earnings. By May 10, 1876, opening day at the Centennial, 859 patents had been granted to women. This number, although large from the point of view of individual women, represents a paltry 0.5 per cent of all patents to that date. But the trend was upwards. The 124 patents granted to women in 1876 alone represent 0.9 per cent of the total for that year. In 1893, the year of the Columbian Exposition in Chicago, where women again put together an impressive pavilion, women obtained 303 patents, or 1.5 per cent of that year's total.[7]

In the 1870s the Patent Office charged a basic fee of $35 and required that a model accompany each application. In theory, an inventor could file his or her own patent application by following the published Rules of Practice, but in practice most inventors hired lawyers to find and argue the patentability of their inventions.[8] Thus the out-of-pocket costs for a patent, to say nothing of an invention, could easily run as high as $100. Women inventors, like those at the Centennial, could seldom afford to seek patents without strong expectations of financial gain.

Seventy-five per cent of the women inventors at the Centennial (59 of the 79) patented their inventions. A few used their patents directly. These, for the most part, were practitioners of pre-industrial crafts--such as corset-making or flower-preserving--and their patents covered improvements in their tools or products.

For most inventors, however, male or female, the patent was the final product. Like a novelist's manuscript, it was exchanged for royalties. The inventor was not necessarily or usually adept at manufacturing or marketing, and even those who were seldom had capital to undertake those lucrative branches of industry. While inventors made some money, they did not often become rich. The practice of selling patents helps explain the difficulty of tracing many inventions. Mrs. Potts' sad-iron was known by that name, and Mrs. Knox' fluting-iron carried her picture. But Mrs. Wells' patented baby-jumper was simply a product of the Occidental Manufacturing Company of Chicago.

A few of the women inventors were sufficiently wealthy that they might have chosen to spend their days as ladies of leisure. Mrs. Anna Weaver of the Woman's Foreign Missionary Society in South America used the profits from her invention to support a school in Bogota. Mary Jackson lived with her mother-in-law, who identified herself as a gentlewoman. But for most of the women the economic realities of nineteenth-century America denied them any opportunity for settling comfortably into a "proper" feminine role or sphere.

Approximately one-fourth of the women inventors at the Centennial were widows or spinsters. The most successful of these was Martha Coston, a woman with little formal education and few marketable skills. At age 22 she was a widow with four small children and no inheritance except some rough notes of her husband Benjamin's unfinished experiments. In short order, she developed these into patentable form, arranged to have her patented pyrotechnic signal flares manufactured and marketed, and signed a most advantageous contract with the Navy, which was then gearing up for the Civil War. In 1876 she was held up as an example of what women could do if the need arose. The Coston Supply Company, which she established, is still in business.

Mrs. Chapman, a Philadelphia corset maker, was a "respectable, industrious woman doing a small retail trade." Unlike Mrs. Coston, she had a husband, but he was an invalid with no outside income. Hannah Supplee, inventor of an easily threaded sewing machine needle, was married to a man "not much interested in business." Mrs. Flynt, a Boston corset maker, was once "quite wealthy, but her husband failed & all of her ppty [sic.], dwindled away." Maria Ware was a florist, her husband was in real estate, and together they made "about a living." Mary Florence Potts, inventor of a truly cold-handled sad-iron, might have fared better. Her young husband was "moral, temperate, prudent, and steady." Nevertheless, in stepping up from clerk to storekeeper he overextended himself and went bankrupt.[9]

In short, these women represent a cross-section of middle-class America. Like other bourgeois groups of the period, these women experienced a large degree of financial instability.[10] For them, invention was not primarily a means of self-expression or a monument to personal vanity, as some have claimed it sometimes was for men,[11] but, rather, a path to the standard American dream of economic self-sufficiency.

Perhaps the best example here of a woman trying to make her own way in the world is Mary Nolan of St. Louis, exhibitor of Nolanum, a patented brick designed to revolutionize building construction. The Centennial judges anticipated "important results" from Nolanum's novel form and kaolin clay composition. Nolanum, Mary's own invention, was but one of many entrepreneurial ventures she undertook after her father's death. In 1870 she advertised herself as a Catholic bookseller. In 1872, assuming editorship of the Inland Monthly Magazine, she wrote, "We have the candor to acknowledge that an improvement of our financial condition, has been one of the motives which induced us to undertake this enterprise." (And she ran an article describing the economic benefits to women of life insurance.) The magazine soon passed into other hands, and neither the millinery shop nor the girls' school Mary and her sisters ran proved more than marginally successful. In 1876, according to a

credit reporter, the whole Nolan family had "great expectations" invested in Nolanum. [12]

The women inventors at the Centennial were intent
on turning their ideas into profit. But many were feminists
as well, aware of the problems and opportunities special to
them as women. The Women's Pavilion as a whole represented a major facet of the nineteenth-century American
woman's struggle for equality, and those women who exhibited
there, rather than elsewhere at the Centennial, chose thereby
to support its aims. The exhibit, and the numerous meetings
and events held in Philadelphia in 1876 on its account,
brought together hundreds of active women, stimulating an
appreciation of their common goals. [13]

The last quarter of the nineteenth century witnessed a
heightened self-consciousness on the part of women inventors.
They knew who they were, as the Patent Office issued a list
of patents granted to women. [14] At the world's fairs held at
Atlanta in 1884-5 and at Chicago in 1893, the women's departments again had large sections devoted to inventions. By
1891 Charlotte Smith, a colleague of the Centennial inventor,
Mary Nolan, was in Washington, publishing a monthly newspaper entitled The Woman Inventor. Among her concerns
were pitfalls that were hazardous for all inventors but apparently particularly so for women: patent fees were un-
reasonably high, some lawyers advocated a lengthy pursuit
of useless patent applications, and ill-advised inventors often
sold their patent rights for ridiculously low prices. Through
the Woman's National Industrial League of America, Smith
worked to improve the situation.

The second and far larger set of industrial problems
were those facing laboring women. Following the well-trod-
den path of reform, the women inventors worked to extend
opportunities within the system as it was, optimistically be-
lieving that America was great enough to provide honorable
jobs at good wages for all needy citizens. One female inventor, a New England clothing manufacturer, insisted on
paying her seamstresses living wages and hoped "never to
do anything to show that to make money we can sacrifice
our principles of justice & right." [15] Many others were
pleased that their inventions provided employment opportunities. The Occidental Manufacturing Company, for instance,
advertised proudly that its product "is wholly the work of a
woman, and that nearly all the work done in the factory is
done by women. "[16] But benevolence was often tempered by

a realistic self-interest. The Emancipation Union Suit, a one-piece flannel undergarment covering torso and limbs, and satisfying all the hygienic principles set forth by the many dress-reform advocates, was patented by Susan Taylor Converse. Manufactured by George Frost & Co., the Union Suit was exhibited at the Centennial, but not in the Women's Pavilion. Female do-gooders wished to lower the price of this garment, to bring its benefits to greater numbers of women. Mrs. Converse replied that she "intended to use any invention for profit," and thus would not forego the twenty-five cent royalty she earned on each garment sold. Indeed, she asked: with all your zeal for women's rights, how could you even suggest that "one woman like myself should give of her head and hand labor without fair compensation?"[17]

The most successful of all women inventors at the Centennial was Ellen Demorest, a couturière with fashion houses in Paris, London and New York. Mme. Demorest, who employed a great many women, might well have taken advantage of those unable to develop from seamstress to entrepreneur, as she herself had done. Instead, however, she lent her support to a variety of innovative self-help programs.

At the launching of the Mme. Demorest, the clipper ship owned by the Women's Tea Company, she had said: "I do not claim that all women, or a large portion of them, should enter into independent business relations with the world; but I do claim that all women should cultivate and respect in themselves an ability to make money as they respect in their fathers, husband, and brothers the same ability." A woman capable of making money "may choose her own position in the world."[18] In summary, then, to make money is mainly what the women inventors at the Centennial were trying to do.

Notes

1. Philadelphia, International Exhibition, 1876, Catalogue of the Women's Department. New Century for Woman, newspaper. Samuel J. Burr, Memorial of the International Exhibition (Hartford, 1877). J. S. Ingram, Centennial Exposition, Described and Illustrated (Philadelphia, 1876). James D. McCabe, Illustrated History of the Centennial Exhibition

(Columbus, 1876). In 1976 the Smithsonian Institution in Washington, D.C. staged a re-creation of the Centennial: see Robert Post, ed., 1876: A Centennial Exhibition (Washington, D.C., 1976), esp. Deborah Jean Warner, "The Women's Pavilion," pp. 163-173.

2. Elizabeth Cady Stanton, S. B. Anthony, & M. J. Gage, History of Woman Suffrage, vol. 3 (Rochester, N.Y., 1887), pp. 29-34, 54-56.

3. Elizabeth Duane Gillespie, A Book of Remembrance (Philadelphia, 1901).

4. See Appendix, following these notes.

5. "Women as Inventors," in The Literary Digest, 1900, vol. 20, pp. 14-15; "Women as Inventors," in Orison S. Marden, ed., The Consolidated Encyclopedic Library (New York, 1903), pp. 4086-4089.

6. "Information Useful to Patentees," with a section on women's rights, Scientific American, 1861, vol. 5, p. 363; "Female Inventive Talent," ibid., 1870, vol. 23, p. 184; "Progress of Woman's Rights," ibid., 1873, vol. 29, p. 7.

7. U.S. Patent Office, Women Inventors to Whom Patents Have Been Granted by the United States Government, 1790 to July 1, 1888 (Washington, D.C., 1888); Appendix #1, July 1, 1888 to October 1, 1892 (Washington, D.C., 1892); Appendix #2, October 1, 1892 to March 1, 1895 (Washington, D.C., 1895).

8. See the patent files, National Records Center, Suitland, Md. Of the 79 women inventors at the Centennial, only one handled her patent application without a lawyer, and one other handled her second application unaided. Because the correspondence was almost exclusively through lawyers, the patent files offer almost no information about the inventors themselves.

9. Most of this information came from the Dun & Bradstreet credit ledgers now housed in the Baker Library of the Harvard University Graduate School of Business Administration.

10. C. Margaret Walsh, "Industrial Opportunity on the Urban Frontier," Wisconsin Magazine of History, 1976, p. 175.

11. Robert C. Post, Physics, Patents & Politics, A Biography of Charles Grafton Page (New York, 1976), p. 106.

12. Cf. Appendix on "Mary Nolan."

13. The New Century Club, the women's club of Philadelphia, was a direct outgrowth of the Women's Pavilion. The Association for the Advancement of Women, the American Women's Suffrage Association, the National Women's Suffrage Association, and the 76 Club (the national association of American women journalists) all met in Philadelphia during the Centennial season.

14. See note 7.

15. Mrs. Horton to Miss Peabody, March 8, 1875. Correspondence, Dress Reform Committee of the New England Women's Club, at Schlesinger Library, Radcliffe College.

16. Quoted in New Century for Women, p. 34.

17. Susan Taylor Converse to Mrs. Wolcott, May 17, 1875. Correspondence, Dress Reform Committee of the New England Women's Club, at Schlesinger Library, Radcliffe College.

18. Quoted in newspaper clipping in Demorest Scrapbook at Schlesinger Library, Radcliffe College.

Appendix

A Reference List of the Women Inventors
at the Centennial

Miss Ellen D. Anderson of Frederick, Md.   Shutter fastener, patent #137,647.

Mrs. Sarah P. Ball (b. 1844) of Philadelphia, and Mrs.
Mary P. Jackson (b. 1837) of Kennett Square, Pa. Sisters.
Gas smoothing iron, patent #177,643. Centennial award.
Wife of Joseph Ball, lawyer; and of Josiah Jackson, busi-
ness manager. Ref.: New Century for Woman, p. 147;
Burr, p. 588; 1870 census.

Mrs. Sarah Hare Bancroft (b. 1824) and Sarah W. Tucker,
both of Media, Pa. Bathing chair for ladies or invalids,
patent #150,510. Shown also at the Franklin Institute in
1874. Wife of Samuel Bancroft, owner of cotton mill. Ref.:
1880 census.

Mrs. Mary Blauvelt (b. 1832) of Ithaca, N.Y. Dressmaker's
marking and cutting gauge, patent #141,849. Wife of William
Blauvelt, blacksmith. Ref.: 1870 census.

Mrs. Maria Bradley of New York. Portable lunch heaters,
patents #168,133 and #195,255. Shown also at Paris in
1878. Ref.: New Century for Woman, #3, p. 1.

C. S. Brooks of Philadelphia. System for cutting clothing.

Mrs. Caroline S. Brooks (b. 1840) of Helena, Ark. Method
of producing lubricated molds in plaster, patent #187,095.
At the Centennial she exhibited the head of the dreaming
Iolanthe modeled in butter. In 1878 she went to Paris, in-
tending to exhibit a full length butter model of Iolanthe; at
Chicago in 1893 she showed a bas relief of George Eliot.
Wife of Samuel Brooks, farmer; daughter of Abel Shawk,
inventor. Ref.: New Century for Woman, pp. 84-85;
Women's Journal, 1877, vol. 8, p. 116, and 1878, vol. 9,
pp. 97 & 273; Burr, p. 591; Ingram, pp. 705-6.

Madame S. C. Brosse of San Francisco. Model for fitting
ladies' dresses by self-measurement. Shown also at the
California State Fair in 1872. Ref.: New Century for
Woman, #3, p. 1; Woman's Words, 1877-79, vol. 1, p. 14.

Mrs. Mary A. Browne of Chicago. Parlor cook stove
adapted to the use of coal. Ref.: New Century for Woman,
p. 114.

Mrs. Mary P. Carpenter of New York. Sewing machine.
Her inventions relating to sewing machines included a needle
and arm, patent #99,158; a feeding mechanism, patent
#112,016; a sewing machine, patent #131,739; button, patent

#137,824; and machine for sewing straw braid, patent #171,774. In 1885, then as Mrs. Mary P. C. Hooper, she invented a grated shovel, patent #316,623; and later a device for numbering houses, patent #355,130; and a netting canopy for beds, patent #364,415.

Mrs. Hariet M. Chapman of Philadelphia. Skirt-supporting shoulder brace and puff corset, patent #172,969; also patent #198,348. Centennial award. Ref.: Dun & Bradstreet credit ledger, Philadelphia, vol. 145, p. 260.

Miss Laura M. Chapman (b. 1838) of Friendship, N.Y. Lap tables, patent #151,090. In 1870 she lived with Hannah Chapman, presumably her mother. Ref.: New Century for Woman, p. 69; 1870 census.

Henrietta H. Cole of Washington, D.C. Fluting machine, patent #55,469. Centennial award. Shown also at American Institute in New York in 1873 (diploma) and in 1874 (bronze medal).

Miss A. V. Coles of New York. Folding cribbage board. Ref.: New Century for Woman, p. 69.

Mrs. Margaret Plunkett Colvin (1828-1894) of Battle Creek, Mich. Triumph rotary washing machine, patent #120,717. Also shown at Chicago in 1893. Other inventions included clothes pounders, patents #199,693, #202,792, and #248,712. Wife of Ashley Colvin (1816-1883), who made and marketed her washing machines. Ref.: 1870 census.

Mrs. Elmira Cornwell of Philadelphia. Graduated chart for dress cutting. Centennial award.

Mrs. Martha J. Coston of Washington, D.C. Pyrotechnic night signals, patents #23,536 and #115,935. Centennial award. Shown also at Paris in 1878, and at Chicago in 1893. Widow of Benjamin F. Coston, U.S.N. Ref.: Martha J. Coston, A Signal Success (Philadelphia, 1886); "Women in Business," in Demorest's Monthly Magazine, 1876, pp. 454-6.

Miss Adelia C. Covell of New York. Perspective outline models for teaching drawing, patent #139,237. Ref.: New Century for Woman, p. 34; Burr, p. 588.

Ellen Louise Curtis Demorest (1825-1898) of New York.

Together with her husband, William Jennings Demorest, she ran a leading international fashion house; they also sold paper patterns and a variety of undergarments, and edited a monthly fashion magazine. Centennial award. Also high premiums from the American Institute in New York, the Massachusetts Charitable Mechanics' Association, and other state fairs, over several decades. Madame Demorest, inventor of the first really excellent cheap hoop skirt and an elegant yet comfortable corset, also devised an automatic floor for elevator shafts, patent #259,105, and a puff for head dresses, patent #265,164. Ref.: New Century for Woman, p. 163; Demorest's Illustrated Monthly Magazine for 1876; Matthew Hale Smith, Successful Folks (Hartford, 1878), pp. 427-31; Notable American Women, 1607-1950, Edward T. James, ed. (Cambridge, Mass., 1971).

Mrs. L. Drury of Springfield, Ohio. Dress-cutting system. Shown also at the Cincinnati Industrial Exhibitions of 1875 and 1879 (premium). Ref.: New Century for Woman, p. 34; Burr, p. 588.

Temperance P. Edson of Dedham, Mass. Self-inflating life preserver. Patent #48,539 refers to inflator for raising sunken vessels. Ref.: New Century for Woman, #3, p. 1.

Mrs. Mary Evard (b. 1825) of Leesburg, Va. Reliance cook stove, patents #76,314, #76,315, and #76,316. Shown also at the Agricultural Exposition at St. Joseph, Mo., in 1873 (silver medal). Wife of Charles Evard, watchmaker from Switzerland. Ref.: New Century for Woman, p. 74; Burr, p. 587; Dun & Bradstreet credit ledger for Virginia, vol. 24, pp. 205, 258x, and 277; 1870 census.

Miss Ellen Eliza Fitz of Boston. Globe, patent #158,581. Ref.: Handbook of the terrestrial globe; or, guide to Fitz's new method of mounting and operating globes (Boston, 1876, 1878, 1880, & 1888).

Mrs. Olivia P. Flynt of Boston. Underclothing for women and children, patents #143,002, #144,609, #146,175, #156,018, #156,019, and #173,611. Centennial award. Garments shown also at the Massachusetts Charitable Mechanics' Association in 1878, at Atlanta in 1884-5, and at Chicago in 1893. Ref.: Mrs. O. P. Flynt, Manual of hygienic modes of under-dressing for women and children (Boston, 1882); New Century for Woman, #3, p. 1, Dun & Bradstreet credit ledgers for Massachusetts, vol. 82, p. 11, and vol. 86, p. 140.

Mrs. Julia Foster of Philadelphia. Bed-dusting rack, patent #178,135. Later inventions included a bill file, patent #181,663, and a needle-threading device, patent #182,572. Ref.: New Century for Woman, p. 69.

Jane Fox of Stamford, N.Y. Dish drainer, patent #160,176.

Mrs. Elizabeth J. French of Philadelphia. Electrotherapeutic appliances, patent #167,162. Shown also at the Philadelphia International Electrical Exhibition in 1884. Widow of Joseph French, civil engineer. Ref.: Elizabeth French, A new path in electrical therapeutics (Philadelphia, 1873, and later editions); her other books on this subject; New Century for Woman, #3, p. 1, and p. 58.

Julia Blanche French of Boston. Bedstead with drawers below, patent #152,357. Centennial award. Later inventions included a combined wash and bath tub, patent #317,333, and a tub or tank for toilet, laundry, and refrigerating purposes, patent #348,321.

Mrs. Ann L. H. Graham (b. 1832) of Chester, Pa. Gridiron for beefsteak with receptacle for gravy, patent #101,361. Wife of Ridgely Graham, physician. Ref.: New Century for Woman, #3, p. 1; 1870 census.

Madame Cathrine A. Griswold of New York. Abdominal skirt-supporting corset, patents #56,210, #61,825, #116,585, #157,445, #171,012, and #181,330, as well as five more obtained after the Centennial. Ref.: advertisements in the Smithsonian Institution Collection of Business Americana.

Mrs. Ellen M. Griswold (b. 1821) of Hagerstown, Md. Support for window sash, patent #61,335. In 1870 she was employed as a schoolteacher, and was probably a widow. Ref.: 1870 census.

Mrs. Elizabeth G. Harley of Haddonfield, N.J. The Complete Darner, patent #156,702. Centennial award. Sold in the Women's Pavilion for 50¢. She later invented a device for attaching umbrellas to clothing, patent #296,002. Ref.: New Century for Woman, p. 69; Ingram, pp. 346-7; Philadelphia city directory for 1877; illustrated advertisement in the Franklin Institute Committee on Science & Arts files.

Emma Heller. Preserve jars. Ref.: Arthur's Home Magazine, 1876, vol. 44, pp. 397-9.

Mrs. D. Grace Hunkins of Allegheny, Pa. Rolling pin holding 10 cooking utensils. Ref.: New Century for Woman, p. 69.

Susan T. Inesly of New York. Sad-iron stand, patent #148,706.

Miss Elizabeth C. Jay of New York. Postage stamp moistener.

Phoebe M. Kelsey of Philadelphia. Meat tenderer. Ref.: New Century for Woman, p. 69.

Mrs. Celine Laumonier of New York, but a citizen of France. Combined traveling bag and chair, patent #168,402. Ref.: New Century for Woman, #3, p. 1.

Lizzie M. Livingston of New York. Garment-cutting system.

Miss Linda H. McNair of Oakland, Cal. Compound-writing instrument, patent #155,196.

Mrs. Carrie Mitchell of Normal, Ill. Excelsior Table and Three-Leaf Hinge. Ref.: New Century for Woman, p. 114.

Mrs. Hannah Mountain of New York. Life-saving mattress, patent #136,749. Shown also at the American Institute in New York in 1873 (diploma). Ref.: New Century for Woman, #3, p. 1; Burr, p. 588; Ingram, p. 345.

Mary Nolan of St. Louis. A new building material called Nolanum, patents #186,604 and #188,660. Centennial award. Ref.: New Century for Woman, p. 115; Burr, pp. 587-8; Inland Monthly Magazine, 1872; Dun & Bradstreet credit ledger, Missouri, vol. 41, pp. 234 & 392.

Mrs. Elizabeth M. Page of Philadelphia. Dirt catcher, a large pan on wheels into which the house maid sweeps. Ref.: New Century for Woman, #3, p. 1.

Miss Emmeline W. Philbrook of Boston. The Equipoise Waist, patent #175,154. Other Philbrook designs for reform undergarments were covered by patents #184,545, #385,570, and #503,436. The Equipoise Waist was sold by dress-reform advocates across the country, and it was advertised in women's magazines at least until the end of the century. Philbrook's other inventions included a spring clasp, patent #177,882; a ventilating strip for windows, patent #286,555; a weather strip, patent #305,471; and a clothes hook, patent #335,237. Ref.: New Century for Woman, pp. 117 & 164.

Mrs. Mary R. Pierce of New York. Improved flower stand. Also shown at the American Institute in New York in 1875 (diploma). Pierce also displayed a thread and needle bank with places for every article used in sewing, and she operated a scroll saw. Ref.: New Century for Woman, p. 69; Ingram, pp. 345-6; Friends' Intelligencer, 1876-7, vol. 33, p. 556.

Mrs. Mary Florence Potts of Philadelphia. Cold-handled sad-iron, patents #103,501 and #113,448. In the 1870s these irons were sold by Hibberd Chalfont and Mary F. Potts, t/a Chalfont & Potts; later they were manufactured and sold by the Enterprise Mfg. Co. and the American Machine Co. With her husband, Joseph H. Potts, she invented another improved sad-iron, patent #506,252, and a remedial or medical appliance, patent #468,946. Ref.: New Century for Woman, p. 147; Burr, p. 588; Dun & Bradstreet credit ledgers for Iowa, vol. 51, pp. 48 & 97.

Miss L. E. Robbins of Boston. Dress-cutting chart. Ref.: New Century for Woman, #3, p. 1; Women's Journal, 1876, vol. 7, p. 112.

Mrs. Clara A. Rogers of New Orleans. Scissors with tools for repairing sewing machines, patent #179,730. Sold in Philadelphia in 1876. Ref.: New Century for Woman, pp. 135-6.

Mrs. Sarah Ruth of Philadelphia. Sunshade for horses, patents #81,412 and #134,564. Ref.: New Century for Woman, #3, p. 1.

Mrs. Elizabeth F. Shaw of Brooklyn, N.Y. Dress protector for sewing machines.

Mrs. Lydia H. Sheppard of Philadelphia. Picture exhibitor, patent #171,585.

Amanda S. Sherwood of Philadelphia. Griddle greaser. Shown also at the Franklin Institute Exhibition in 1874 (honorable mention). Ref.: New Century for Woman, #3, p. 1; Arthur's Home Magazine, 1876, vol. 44, pp. 450-2.

Mrs. Robert Shields of Chicago. Eccentric gauge used to make candy lozenges. Ref.: Burr, pp. 585 & 587.

Mrs. Semele Short of Cincinnati, Ohio. Blanket washer, mangle, and drying frame for curtains, covered by patents #145,910, #145,911, and #145,912. Centennial award. Also shown at the Cincinnati Industrial Expositions of 1874 (silver medal, bronze medal, and premium), of 1879 (silver medal, bronze medal, and premium), and of 1883.

Mrs. Charlotte L. Slade of New York. Combination desk and slate, patents #177, 892 and #179,362. An improvement in dolls' hats received patent #193,674. C. L. Slade & Co. exhibited dolls' clothes at the American Institute in New York in 1873, and the Slade Manf'g and Publishing Co. showed dolls' patterns and such there in 1875 (diploma). Widow of Lloyd Slade. Ref.: New Century for Woman, p. 212; New York City directory for 1876; Dun & Bradstreet credit ledger for New York City, vol. 247, p. 2381.

Mrs. Martha E. Slocum of New York. Plant protector.

Mrs. Orrin Smith of Chicago. Cooking range, patent #54,427. Wife of Orrin Smith, commission merchant. Ref.: New Century for Woman, p. 114.

Mrs. Jennie H. Spofford of Philadelphia. Mattress supporter, mosquito net frame, and spring saddle, patents #176,413, #176,370, and #176,206.

Mrs. B. A. Stearns of Woburn, Mass. Dress-cutting chart. Centennial award. Shown also at the Massachusetts Charitable Mechanics' Association in 1869; the American Institute in New York in 1877 (medal of excellence); and at Atlanta in 1884-5. Ref.: New Century for Woman, p. 53; Burr, p. 588.

Mrs. Hannah Steiger of Laurel, Md. Locking barrel cover, patent #42,699. Wife of William T. Steiger, lawyer. Ref.: New Century for Woman, p. 69.

Mrs. Charlotte Higgins Sterling of Gambier, Ohio. Dish washer, patent #130,761. Centennial award. Wife of Theodore Sterling (1827-1912), professor of various sciences and later president of Kenyon College. Ref.: New Century for Woman, #3, p. 1; Arthur's Home Magazine, 1876, vol. 44, pp. 450-2.

Mrs. Elizabeth Mary Stigale of Philadelphia. A model for cemetery lots, with ornamental recessed rail for the growth of flowers, covered by patents #97,984 and #112,748. Her methods of preserving flowers were covered by patents #84,445 and #89,515. Widow of Robert Stigale, bricklayer. Ref.: New Century for Woman, #3, p. 1; Mrs. Stigale's circular for natural flowers restored and preserved, in her patent files.

Mrs. Elizabeth W. Stiles of Philadelphia. Combination desk, patent #167,586. Centennial award. Shown also at the American Institute in New York in 1875 (silver medal). Mrs. Stiles also exhibited a binder for paper files, patent #178,975, and a revolving ink stand, patent #195,462. Ref.: New Century for Woman, pp. 51 & 204; Burr, pp. 586-7; illustrated advertisement in Franklin Institute Committee on Science and Art files.

Mrs. Hannah Suplee of Philadelphia. Open-eyed sewing machine needle, patent #94,924. Shown also at the American Institute in New York in 1875 (bronze medal) and in 1876 (diploma of continued excellence). Suplee also invented a pattern and lining for garments, patent #250,998, and an abdominal supporter, patent #500,356. Wife of Albert H. Suplee, agent for sewing machines. Ref.: New Century for Woman, p. 91; Ingram, pp. 368-9; Woman's Journal, 1875, vol. 6, p. 299; Dun & Bradstreet credit ledgers for San Francisco, vol. 14, p. 165, and vol. 16, p. 267, and for New York, vol. 245, p. 2164.

Mrs. Georgiana L. Townsend of Philadelphia. Vertical handle attachment for sewing machines, patent #177,084. A member of the Women's Centennial Committee, Mrs. Townsend was chairman of the space allocation group. Wife of Henry C. Townsend, lawyer. Ref.: New Century for Woman, p. 91; Ingram, p. 368.

Miss Marietta Tremper of New York. Window-washing machine.

Catharine L. Tresize of Springfield, Ill. Trunk, patent #175,208. Ref.: New Century for Woman, p. 34.

Annie C. Vogel of New York and Fannie M. Krebs of Georgetown, D.C. Hair-curling pin, patent #102,338.

Mrs. Maria L. Ware of Philadelphia. Natural flowers

preserved by a chemical process, patent #134,714. Centennial award. Wife of John W. Ware, realtor. Ref.: New Century for Woman, #2, p. 2; Dun & Bradstreet credit ledger for Philadelphia, vol. 151, p. 138.

Mrs. Anna K. Weaver of Salem, Ohio. Photographs of chaste and beautiful mottos in fern leaves. Wife of Rev. Willis Weaver. Ref.: New Century for Woman, p. 180; Burr, p. 638; Friends' Intelligencer, 1876-7, vol. 33, p. 556.

Miss Glory Ann Wells of Luzerne, N.Y. Dish washer, patent #104,235.

Mrs. Jane E. Wells of Chicago. Baby jumper, patent #130,397. Produced by the Occidental Mfg. Co. of Chicago, of which her husband, Joel Henry Wells, was manager. Shown also at the Cincinnati Industrial Exposition of 1874. Ref.: New Century for Woman, pp. 34 & 114.

Mrs. Emma J. Whitman (b. 1845) of Oakland, Cal. Kettle and pan scraper, patent #142,602. Wife of William W. Whitman, grocer. Ref.: 1880 census.

Mrs. Mary A. E. Whitner of Philadelphia. Stereoscope, patent #148,555.

Mrs. Anne C. Wilhelm of Philadelphia. Button for garments, patent #135,739. After her husband Charles died, she succeeded him as a partner in the firm of Wilhelm & Neumann, makers of lamps and other railroad supplies. Ref.: Dun & Bradstreet credit ledgers for Philadelphia, vol. 134, p. 461, and vol. 158, p. 305.

Mrs. G. A. Williams of Baltimore. Leather table with checker board.

Mrs. Caroline Wimpfheimer of Philadelphia. Loom for making hair ribbon. In 1876 she was employing 60 hands in her business. Widowed. Ref.: Dun & Bradstreet credit ledger for Philadelphia, vol. 149, p. 241.

Miss L. Frances Woodward of Woodstock, Vt. Ladies' work table, patent #177,779.

2. WOMEN SCIENTISTS IN AMERICA BEFORE 1920*

by Margaret W. Rossiter

If asked to name a woman scientist, most Americans would probably mention Marie Curie, the French woman famed for her researches on radioactivity and her two Nobel Prizes. If pressed to name an American woman scientist, most would probably refer to Ruth Benedict, Margaret Mead, or Rachel Carson, whose books have gone through numerous editions in recent decades. But few would know of any other women scientists. This lack of awareness has, unfortunately, led to two widespread but incorrect impressions: (1) women scientists are isolated phenomena and (2) it is only very recently that women in America have begun to take up careers in science. It may come as a surprise, therefore, to learn that women have been a part of the American scientific community for over a century, and that, far from being isolated phenomena, they constituted a sizable group of strong-willed individuals, many of whom worked hard to overcome those very problems of rejection and discrimination that have become an object of concern in recent years.

When in 1906 James McKeen Cattell, editor of Science and professor of psychology at Columbia University, published the first edition of his American Men of Science: A Biographical Directory, he included 149 women on his list of 4,131 scientists (3.6 per cent). His second edition, in 1910, contained 204 women, but their percentage of the total dropped slightly to 3.5 per cent. By 1921, after a decade of rapid growth, the number had more than doubled to 459, but, since the number of male scientists was also increasing, the percentage of women rose only to 4.8 per

*Reprinted by permission from American Scientist (journal of Sigma Xi, The Scientific Research Society of North America), Vol. 62, No. 3, May-June, 1974, pp. 312-323. Copyright © 1974 by The Society of the Sigma Xi.

cent.   Together these first three editions contained 504 women who were seriously interested in science and wished to pursue careers in it before 1920.[1]

Although we know very little about the early women of science, these directories provide a fairly complete curriculum vitae for each individual and attempt to assess the quality of her work.   They thus provide one way to study the career patterns of a large number of women scientists before 1920.   Their situation is of contemporary interest, for many of the current problems of women scientists were present before 1920.   Although their interests and backgrounds varied widely, they clustered in certain fields and certain institutions, and, despite an average level of education higher than that of the male scientists, they had fewer job opportunities and lower status, were more often unemployed, and less often considered eminent by their fellow scientists.   A few women managed to overcome the pattern of discrimination and do important work.   Others formed organizations to help correct the unequal conditions and to enhance the status of their sex in science.   The story of these 500 women and their efforts to pursue scientific careers provides an interesting glimpse into a neglected aspect of the history of science.

Male Patterns

Before we can understand the place of women scientists in the scientific professions at the turn of the century, we must have some picture of the number, distribution, and career patterns of the men.   Although governments did not keep or publish accurate statistics on men and women of science until long after 1900, we can perhaps get a rough picture of the career patterns of scientific men from a sample of 502 chosen from the third edition of American Men of Science (1921).[2]

Table 1 summarizes professional biographical data on these men.   The largest group are in chemistry, with 17.9 per cent of the total.   The second largest group are in the medical sciences of anatomy, bacteriology, physiology, and pathology.   Together these two groups comprise about one-third of the total.   Ranking next are four fields of about equal size--botany, engineering, physics, and zoology (including entomology)--which together constitute another 40 per cent of the male scientists.   All other scientists are divided among seven smaller fields. The reasons for this particular distribution of scientists must be exceedingly complex, the result of many economic, social, intellectual, and personal factors. The most that can be said is that such a distribution somehow reflected the interests and needs of industrial America at the turn of the century.

The educational patterns within each field were likewise quite varied. As late as 1920 men could still enter the scientific profession without a higher degree, but by that year a high percentage (46.6 per cent) held the Ph.D. Psychology, one of the smaller sciences, had the highest percentage of doctorates (92.3 per cent), which perhaps reflects the fact that it, more than any other science, grew up with the American university. The Ph.D. was the least common in pharmacy, engineering, agricultural sciences, and medical sciences, where it had not yet replaced the M.D. But in most sciences the usual pattern was a mix of one-third to three-fourths younger men with doctorates and the rest older men without.

Table 1 provides some indication of the types and variety of jobs held by men in each field. The picture can only be approximate since job mobility was high and many people held more than one job at a time. However, the table gives some idea of the relative importance of academic and nonacademic employment in the various sciences. Once again, the pattern was different in each field. By 1920, most of the scientists (63.1 per cent) were employed in a university or in an affiliated institution such as a museum, observatory, or research institute. The most "academic" of the sciences--those with the smallest percentage employed in government or industry--were mathematics and psychology; the most "applied" were chemistry, entomology, and engineering, which had extensive industrial as well as governmental applications. The other sciences had applications which were primarily in government.

Doctorates were highly correlated with academic positions. If one lists the sciences in the order of their decreasing dependence on academia for jobs, the order is roughly the same as that of the percentage of doctorates in each field (Table 2). The exceptions are pharmacy and medical sciences, where only the elite practitioners were included in the AMS directories, and chemistry, where a doctorate was often necessary even for work in industry. Although it is not possible to include all the men's occupations

Table 1. Fields, doctorates, and occupations of a sample of male scientists.

Field	No. men	% of total	Ph.D.'s	% holding Ph.D.'s	Jobs in academia		% in academic institutions	Other occupations (more than 1)
					Profs.	Museums etc.		
Chemistry	90	17.9	59	65.6	36	0	40.0	Industrial chemist, 24; govt. chemist, 10; executive, 8; consulting chemist, 4; hospital chemist, 3; editor, 2; teacher, 2.
Medical sciences	77	15.3	17 (59 M.D.)	22.1	59	1	77.9	Private or hospital physician, 11; govt. bacteriologist, 2; public health, 2.
Botany	49	9.8	20	40.8	29	0	59.2	Govt. botanist, 10; govt. forester, 3.
Engineering	49	9.8	11	22.4	15	0	30.6	Consulting engineer, 13; govt. engineer, 8; industrial engineer, 5; industrial executive, 5.
Physics	48	9.6	34	70.8	39	0	81.3	Govt. physicist, 8.
Mathematics	38	7.6	28	73.7	35	0	92.1	—
Zoology	37	7.4	21	56.8	20	8	75.7	Govt. biologist, 3.
Geology	35	7.0	12	34.3	18	7	71.4	Govt. geologist, 7; petroleum geologist, 3.
Agricultural sciences	23	4.6	7	30.4	13	0	56.5	Govt. agronomist, 4; govt. husbandryman, 4.
Astronomy	17	3.4	7	41.2	5	7	70.6	—
Entomology	15	3.0	2	13.3	3	3	40.0	Govt. entomologist, 3.
Psychology	13	2.6	12	92.3	11	0	84.6	—
Anthropology	6	1.2	3	50.0	3	1	66.7	—
Pharmacy	5	1.0	1	20.0	4	0	80.0	—
Total	502	100.2	234	46.6 (avg.)	317		63.1 (avg.)	

Source: Third edition of American Men of Science, 1921.

in Table 1, a quick scanning of
the most popular ones shows
that male scientists permeated
the major areas of the American economy from mining to
agriculture to industry and
medicine. There seemed to
be few limits on where an
American scientist--male that
is--might appear.

Female Patterns

Table 3 provides a sum-
mary of the biographical data
on the 504 women scientists in
the first three editions of <u>American Men of Science</u>.[3] The
most popular sciences were zoology, botany, and psychology,
which together attracted over half the total group. The rest
of the women were spread over eight smaller fields. Anthropology was, as with the men, one of the smallest sciences,
since its growth did not occur until after 1920. Quite striking was the total absence of women from engineering (which
was so popular with men), pharmacy, and the agricultural
sciences of animal nutrition and agronomy.

The reasons for this particular distribution of women
scientists are unclear. Most probably lacked rigorous mathematical training, as Alice Hamilton, an eminent industrial
toxicologist, admitted in her autobiography,[4] although the
minority in the physical sciences excelled in it. Perhaps
the early women's colleges, which trained and employed
many of these women, may have stressed some sciences
over others, as Mount Holyoke did zoology (see Table 4).
Or perhaps women were counseled in some other way to
prefer zoology, botany, and psychology.

It is equally difficult to determine why women did not
enter some of the other sciences. Since there was no woman
engineer in this period [in this sample], it is hard to tell how one
might have fared. Would she have been permitted to enroll in an
engineering school, if she had been attracted to such a "masculine" field in the first place? Or would she have been re-
pulsed by the hurly-burly of a field that offered relatively
few academic positions (and none in the women's colleges)
and which might require rigorous outdoor work? Such self-

Table 2. Percentage of male scientists
employed in academic positions, 1920.

	%
Mathematics	92.1
Psychology	84.6
Physics	81.3
Pharmacy	80.0
Medical sciences	77.9
Zoology	75.7
Geology	71.4
Astronomy	70.6
Anthropology	66.7
Botany	59.2
Agricultural sciences	56.5
Chemistry	40.0
Entomology	40.0
Engineering	30.6

Source: Sample from the 3rd edition of
American Men of Science.

selection was very likely the case, since in general women were found less often in those fields with the most industrial and governmental applications: engineering, entomology, and chemistry.

If rejection or a fear of rejection kept women out of some fields, another force was attracting them into the new "female" science of home economics around 1900-1910. Home economics was the result of a sexual division of scientific labor--the men took animal nutrition and the women human nutrition. The 23 women who identified themselves as being in the science of home economics before 1920 were a very well-educated group. Seventeen held Ph. D.'s, usually in chemistry and especially in the area known as "physiological chemistry," from which the modern sciences of biochemistry, pharmacology, nutrition, and microbiology have all emerged. Several women studied food chemistry with Wilbur O. Atwater at Wesleyan University, but the major graduate programs in this field at the time were those of Russell H. Chittenden and Lafayette B. Mendel at Yale University, where 7 of the 17 earned their degrees, and the University of Chicago, with Julius Stieglitz and Marion Talbot, where 4 got Ph. D.'s.

An examination of the subsequent careers of the Yale doctorates in physiological chemistry reveals that most of the women entered home economics and most of the men biochemistry, pharmacology, and animal nutrition.[5] What was happening was that newly created jobs in home economics at the large state universities attracted women, who were not as readily employed in biochemistry or animal nutrition as were men. For example, Isabel Bevier, one of Atwater's students, was teaching chemistry at Lake Erie College in 1899 when she was called to the University of Illinois to head one of the first home economics departments. Likewise, Mary Swartz Rose, Louise Stanley, Amy Daniels, and Ruth Wheeler were all chemists who were able to move into the new positions in home economics which were created after 1900. This whole home-economics movement owed much to the efforts of Ellen H. Richards (Fig. 1), lecturer on sanitary chemistry at M. I. T., who had been popularizing the subject since the 1880s. By 1910 home economics had been accepted as a woman's science, although the founders of the parent disciplines of food and physiological chemistry, from Justus Liebig on, had all been men.[6]

Table 3 also shows that, on the whole, the women

Table 3. Fields, doctorates, marital status, and occupations of women scientists.

Field	No.	% of total	Ph.D.'s	% holding Ph.D.'s	Married	% married	Jobs in academia Profs.	Other*	% in academia	Other occupations (more than 1)
Zoology	92	18.3	59	64.1	26	28.3	50	9	64.1	†Unemployed, 11; high school teacher, 7; ‡unknown, 7.
Botany	91	18.1	49	53.8	17	18.7	50	6	61.5	Govt. botanist, 24; unknown, 6; high school teacher, 3; unemployed, 2.
Psychology	69	13.7	62	89.9	18	26.1	40	4	63.8	School psychologist, 9; psychologist for social agency, 8; unemployed, 4; research ass't, 3.
Medical sciences	63	12.5	24 (37 M.D.)	38.1	13	20.6	26	9	55.6	Govt. public health work, 7; private physician, 6; unknown, 6; research ass't, 6; hospital staff physician, 4; curator, 4; editor, 2.
Mathematics	46	9.1	36	78.3	6	13.0	40	—	87.0	Unknown, 2.
Chemistry	42	8.3	34	81.0	6	14.3	28	—	66.7	Unemployed, 3; unknown, 2.
Geology	25	5.0	16	64.0	3	12.0	14	7	84.0	—
Home economics	23	4.6	17	73.9	3	13.0	18	—	78.3	USDA home economist, 2.
Physics	23	4.6	15	65.2	0	0.0	18	—	78.3	High school teacher, 3.
Astronomy	21	4.2	5	23.8	2	9.5	10	10	95.2	—
Anthropology	9	1.8	2	22.2	4	44.4	1	4	55.6	Unknown, 2.
Total	504	100.2	319	63.3 (avg.)	98	19.4 (avg.)	344		68.3 (avg.)	

*Museums and observatories.
†Unemployed implies having once held a job or earned a degree.
‡Unknown indicates probably an amateur, listing neither a degree nor a job.

Source: First three editions of *American Men of Science*.

Table 4. Baccalaureate origins of American women of science before 1920. Only those colleges and universities with more than seven graduates are listed.

	Total	Zool-ogy	Botany	Psychol-ogy	Medical sciences	Mathe-matics	Chem-istry	Home econom-ics	Physics	Astron-omy	Geol-ogy	An-thro-pol-ogy
Wellesley	36	2	6	4	7	8	3	—	4	1	1	—
Vassar	35	6	1	6	2	4	4	4	1	5	2	—
Smith	29	5	5	5	2	5	3	1	—	3	—	—
Mount Holyoke	28	9	1	4	2	3	5	1	2	—	1	—
Cornell	23	5	4	3	3	2	1	1	2	—	2	—
U. of Michigan	21	7	5	4	—	—	2	2	—	1	—	—
Bryn Mawr	19	4	1	1	3	1	3	—	2	—	4	—
U. of Chicago	18	3	5	3	2	—	2	2	—	—	1	—
Barnard	17	1	2	5	1	1	1	2	1	—	1	2
U. of Pennsylvania	12	4	3	2	1	—	2	—	1	—	1	—
Goucher	12	5	1	—	2	1	2	—	—	—	—	—
U. of Nebraska	10	—	4	5	—	—	1	—	—	—	—	—
U. of Wisconsin	9	2	3	1	—	1	1	—	—	—	1	—
Stanford	9	2	3	—	2	—	—	2	—	—	—	—
Oberlin	9	1	4	1	—	2	—	—	1	—	—	—
Radcliffe	8	2	1	1	—	1	—	—	—	2	1	—
U. of Kansas	8	5	—	1	1	—	—	—	—	—	1	—
Ohio State	8	1	4	—	—	—	—	—	—	—	3	—

Source: First three editions of *American Men of Science*.

Figure 1. Ellen Richards, a "sanitary chemist," testing the water of Jamaica Pond, in Massachusetts, about 1898. She taught at M.I.T. and founded the field of home economics. (Permission of The Sophia Smith Collection--Women's History Archive, Smith College, Northampton, Mass.)

scientists before 1920 were a more highly educated group than the men, with 63.3 per cent holding Ph.D.'s compared to 46.6 per cent for the men. (This is in contrast to recent figures, based on the 1960 census, which show men more likely to have doctorates than women.)[7] The women's advantage was not true in all fields, however, and seems due primarily to the low percentage of male doctorates among engineers and medical scientists. Women doctorates predominated in zoology, botany, medicine, mathematics, chemistry, and geology, and men in physics, psychology, astronomy, and anthropology. There seems to be no pattern in this. Nor for women is the percentage of doctorates in a field as closely related to the percentage of jobs in that field in academia as was true for the men. Both percentages were relatively uniform, probably reflecting the limited range of opportunities open to women in all fields.

Well over half the women, like the men, found employment in academic institutions (68.3 per cent for the women vs. 63.1 per cent for the men). The women in academia were more often than the men in positions of inferior status. Almost twice as many (9.7 per cent vs. 5.4 per cent) worked in museums and research institutions rather than in colleges and universities. When they were in such schools, it was more likely to be in the women's colleges, normal colleges, and nursing schools rather than in the universities and medical schools, where almost all the academic men in the sample were employed. This finding is corroborated by a report of the American Association of University Professors, which studied the status of women in academic institutions in 1921. Women at that time held 0.001 per cent of the professorships at men's colleges and universities (2 of "nearly 2,000"), 68 per cent of these at the women's colleges (415 of 613), and 4 per cent of those at the coeducational colleges and universities (190 of 4,760), where they were heavily concentrated in home economics and physical education. This report also found that the women hovered in the ranks of assistant, instructor, and assistant professor at the co-ed institutions far longer than did the men. [8]

However difficult women found careers in academia, outside it they fared even worse. Here their employment picture diverges sharply from that of the men, for non-academic opportunities were quite restricted. Unlike the men, they found no jobs in industry before World War I. The war seems to have opened opportunities to a few women chemists, for two went to work for the National Aniline and Chemical Company during the war and four others took up similar employment between 1918 and 1920. Most notable of this group was Mary Engle Pennington (1872-1952), who became interested in the problem of food spoilage as chief of the Food Research Laboratory of the U.S. Department of Agriculture, and then moved to New York City in 1919, where, as a consulting chemist, she designed commercial and household refrigerators. [9]

The chief opportunities open to women outside

academia were in private practice for the physicians, in
schools and clinics for the psychologists, and in government
bureaus for the botanists.   But not all government agencies
were alike.   One agency, the Bureau of Plant Industry in
the Department of Agriculture, hired the strikingly large
number of 24 women between 1900 and 1920.   This bureau,
like many others, underwent tremendous growth in these
twenty years, but for some unknown reason it alone attracted
a great many women.   Since government salaries were much
higher than those in the women's colleges,[10] women would
probably have entered other government agencies, such as
the National Bureau of Standards, the Forest Service, the
Geological Survey, and other bureaus in the Department of
Agriculture, had the opportunity been open to them.   Each
of these agencies had numerous male scientists, but at most
one or two and usually no women scientists.   This pattern
makes the Bureau of Plant Industry all the more unusual.

The women were also much more likely than the men
to be unemployed or to be employed in jobs unrelated to
science.   This problem had been minor for the men, since
of the 502 in the sample only ten (2.0 per cent) might be
considered to be in this category: four were merchants or
booksellers, two clergymen or missionaries, and four of unknown occupation.   They were spread across six sciences,
but five were in zoology, where a real job shortage may
have existed.   The women zoologists also found unemployment especially high: 21 of 92 (or 22.8 per cent) were
either unemployed or employed outside science as, for example, a missionary, librarian, or archivist.   Several other
zoologists were employed in informal positions as their husbands' assistants.

Marriage, especially high among the zoologists
(Table 3), seems to have been the chief cause of the widespread unemployment in zoology, since 13 of the 21 unemployed were married (61.9 per cent)--more than double the
percentage for the field as a whole (28.3 per cent).   The
career patterns of these women also show sudden breaks and
extended gaps, which, since many were married to other
zoologists listed in American Men of Science--e.g. Gertrude
Crotty Davenport (Mrs. Charles B.), Florence Merriam
Bailey (Mrs. Vernon), Mildred Hoge Richards (Mrs. Aute),
and numerous others--can be explained in terms of their
husbands' careers.   Male zoologists are well known for
marrying a disproportionately large number of distinguished
women, many of them fellow students.[11]

Figure 2. Maria Mitchell, an astronomer, in the dome of the observatory at Vassar College, 1878. She earned fame for her discovery of a comet in 1847. (Courtesy, Vassar College Archives.)

There is also evidence that marriage to a fellow zoologist had a significant effect on a man's career. Of the 26 women zoologists, 15 were married to men in American Men of Science. Of these 15 men, 14 were in zoology or medical sciences, and 8 of them were "starred," or considered eminent by their colleagues, a far higher percentage (57.1 per cent) than was true of zoologists in general (roughly 33.3 per cent in the first edition and 11.1 per cent in the third). In only 3 of these 14 cases did the woman list herself as employed as her husband's assistant, but the rest must also have contributed something to their husbands' careers. Susanna Phelps Gage (Mrs. Simon H.), an embryologist married to a starred histologist/embryologist, was herself starred, although officially unemployed. It is hard to tell what such "unemployment" really meant, if the woman had access to the field and was able to use her talents through her husband's career. It is also hard to tell whether the married woman's unemployment was voluntary or involuntary--the result of social prejudice against the employment of married women, the difficulties of relocation, or the desire to be at home with her children. We can only conclude that these factors, plus a possible job shortage in zoology, resulted in an exceptionally high rate of disguised unemployment for women in zoology before 1920.

In other fields this phenomenon was not so widespread. The rates of both marriage and unemployment were lower. Botany, psychology, and medical sciences--the other fields which contained ten or more married women--were also the fields which, as mentioned above, offered women the widest range of professional employment. In other fields, the percentage of married women was strikingly low, actually reaching zero in physics and approaching it in astronomy, where one and perhaps both of the married women were widows who had to work for a living. The anthropologists, on the other hand, were frequently wealthy widows who were interested in traveling in different cultures and studying the native peoples.[12] Allowing for the fact that American Men of Science is not wholly reliable on the marital status of its women, marriage does seem to have been a significant factor in the employment patterns of women scientists, but in different ways in different fields.

There were thus in 1920 a multitude of patterns and subpatterns in the distribution of women scientists in America. How had such patterns arisen? Were these women really a part of their profession? And were any of the 500 eminent? I can only touch on the answers to these complex questions briefly in the space allotted here.

Women's Education

Jane Colden (1724-1766), the botanist daughter of a famous botanist father, was probably the first woman scientist in America. She had few followers until the second half of the nineteenth century, when important changes in women's education took place. Opportunities for women to study science, previously available only in a few colleges in the Midwest, increased greatly with the founding of the women's colleges and the expansion of the land-grant universities after the Civil War. The relative importance of the women's colleges in training women scientists before 1920 can be seen in Table 4. Of the 459 women who attended one undergraduate college and the 24 who attended two, making a total of

483, 184 (or 38.1 per cent) attended just 8 Eastern women's colleges and 91 (or 18.8 per cent), 7 top state universities, as listed in Table 4. The remaining 208 (42.8 per cent) attended 98 other institutions, the most important of which were the universities of Chicago and Pennsylvania, Stanford University, and Oberlin College.

Women scientists were also dependent on the women's colleges for the large number of jobs their faculties provided. Although each college could provide only a few positions in each science, together they could employ 20-30 women, which was roughly the number of women in some of the smaller sciences. In physics the women's dependence on these colleges reached an extreme. In the years 1900-1920, 21 of the 23 women physicists were at some time employed in women's colleges (91.3 per cent). In the larger sciences the women's colleges played a proportionally smaller role, but even there these professorships were quite prestigious and usually held by the top women in the field, such as Margaret Washburn in psychology at Vassar or Margaret Ferguson in botany at Wellesley.

The number of jobs at the women's colleges more than doubled between 1906 and 1921 (from 43 to 96), as the older colleges expanded and some newer ones--Connecticut College for Women, Sweet Briar, and Skidmore--were established. The overall importance of the women's colleges, however, began to decline after 1906, as more and more opportunities for women opened up in state universities and other schools. Table 5 shows the period's overall educational expansion, which resulted in a steady decline in the percentage of women employed in the women's colleges, from 57.3 per cent in 1906 to 36.5 per cent in 1921. The state universities showed the largest overall increase, jumping from two jobs in 1906 (2.7 per cent) to 53 in 1921 (20.2 per cent). Although the women's colleges were by 1920 still the largest employers of women scientists, their overall dominance had diminished to slightly over one-third of the field, a percentage roughly equivalent to that of their (undergraduate) alumnae among women scientists as a whole. In the period before 1920, the women's colleges thus accounted for roughly one-third of the "supply" and one-third of the "demand" for women scientists.

Although in their heyday of 1905-1920, the faculties of the women's colleges included a number of highly educated and well-trained women, this had not always been the case.

Table 5. Academic institutions employing women, 1906–1921.

1906		1910		1921			
Wellesley	14	Wellesley	12	Wellesley	22	Columbia Teachers	4
Mount Holyoke	8	Mount Holyoke	8	Vassar	15	Connecticut Coll.	4
Bryn Mawr	5	Vassar	6	Mount Holyoke	13	Ohio State	4
Vassar	4	U. of Chicago	6	Smith	13	Sweet Briar	4
Barnard	3	Bryn Mawr	5	U. of California	11	Harvard	4
Smith	3	Smith	5	U. of Chicago	10	Goucher	3
U. of Chicago	3	Barnard	4	Barnard	8	U. of Illinois	3
Goucher	2	U. of Illinois	4	Bryn Mawr	7	U. of Nebraska	3
Sophie Newcomb	2	Goucher	3	U. of Minnesota	7	Stanford	3
Johns Hopkins	2	Johns Hopkins	3	Johns Hopkins	6	Rhode Island Coll.	3
U. of Minnesota	2	Harvard	3	U. of Wisconsin	5	Kansas St. U.	3
Stanford	2	U. of Nebraska	3	Columbia	5	U. of Calif., L.A.	3
Rockford	2	Western Coll.	3	Sophie Newcomb	4	Iowa State	3
N.Y. St. Normal	2			U. of Kansas	4	Carnegie Tech.	3
				Cornell	4	Amer. Coll. for Girls in Istanbul	3
+ 21 others with one each		+ 7 others with 2 each +28 others with 1 each		+ 18 others with 2 each + 43 with 1 each			
Total	75		107		263		

Source: First three editions of *American Men of Science.*

When Matthew Vassar started the movement for the higher education of women by founding Vassar College in Poughkeepsie, New York, in 1862, he sought a woman scientist for his faculty. In 1865 he hired Maria Mitchell (Fig. 2), a Quaker astronomer from Nantucket Island, who had earned fame in 1847 as the first to sight a new comet. She had had no college or seminary training and, in the tradition of Jane Colden, had learned her science by helping her father run his small observatory on the island. During the 1850s she earned her living by working part-time as a piece-work "computer" for the National Almanac Office of the U. S. Coast and Geodetic Survey. At Vassar, Miss Mitchell was a forceful teacher, and several of her students became scientists.[13] Likewise at the other women's colleges established before 1885 (Mount Holyoke, Smith, and Wellesley), the first faculties contained women with only modest levels of education, and many were self-taught.

The big change began in 1885 when M. Carey Thomas of Bryn Mawr College, an advocate of the ideals of Daniel Coit Gilman of The Johns Hopkins University (although he would not allow her to attend its classes), ambitiously set out to hire women with advanced degrees and scholarly publications for her new college and graduate school. By the mid-1880s a few such women were available, and over the years Miss Thomas managed to hire some of the most outstanding women scientists of her time: for example, Charlotte Scott in mathematics, Florence Bascom in geology, and Nettie Stevens in zoology. Following Miss Thomas's example, the women's colleges, always anxious to prove themselves equal to the men's colleges, rapidly transformed their faculties into the multidegreed ones they became by 1910. The pressure to have advanced degrees became so great that even women who had become professors in an earlier day when degrees did not matter, such as Cornelia Clapp of Mount Holyoke (Fig. 3), took leaves of absence and went off to earn Ph.D.'s with their own former students.

The process of coeducating the graduate schools in the 1880s and 1890s was a difficult and painful one. Most graduate schools had no policy regarding women, and when the first hardy souls presented themselves, the schools did not know what to do. Christine Ladd (later Ladd-Franklin) was one of the early graduate students at The Johns Hopkins University, but when she presented her thesis to the mathematics department in 1882, she discovered that her degree would be withheld. It was finally awarded, over 40 years later, in 1926. Florence Bascom, daughter of the president of the University of Wisconsin, was more fortunate: she obtained her Ph.D. in geology at Johns Hopkins in 1893 by special dispensation, although women were not officially admitted until 1907.

At Harvard, Mary Whiton Calkins studied psychology and wrote a thesis with William James and Hugo Münsterberg, but discovered in 1896 that the Harvard Corporation would not grant her a degree. Luckily this injustice proved no lasting obstacle to Miss Calkins, who became a professor at Wellesley and one of the early presidents of the American Psychological Association. In 1902 Harvard moved to clarify the situation by creating the separate Radcliffe Graduate School, which would grant higher degrees to women at Harvard, but the school granted very few degrees in the sciences before 1920.[14]

Table 6. Institutions which granted 5 or more Ph.D.'s to women before 1920, by field.

	Total	Anthropology	Astronomy	Botany	Chemistry	Home economics*	Geology	Mathematics	Medical sciences	Physics	Psychology	Zoology
U. of Chicago	63	—	1	18	5	4	1	5	5	2	15	7
Cornell	34	—	—	4	4	1	2	3	2	3	10	5
Columbia	30	1	2	2	1	—	3	2	2	—	9	8
Bryn Mawr	27	—	—	—	7	1	3	4	1	2	2	7
U. of Pennsylvania	24	1	—	2	5	—	—	2	3	2	3	6
Yale	22	—	—	1	3	7	1	6	1	—	1	3
Johns Hopkins	15	—	—	1	2	1	2	2	1	2	3	1
U. of California	10	—	—	2	—	—	—	—	2	—	2	4
U. of Michigan	9	—	1	—	2	—	1	2	1	—	—	2
U. of Illinois	8	—	—	—	1	1	—	2	—	—	2	2
U. of Wisconsin	8	—	—	4	1	—	—	—	1	—	—	2
U. of Zurich	8	—	—	2	1	—	—	—	—	—	1	4
Clark	6	—	—	—	—	—	—	—	1	1	4	—
Brown	5	—	—	1	—	—	2	—	1	—	1	1
Indiana	5	—	1	2	—	—	1	1	—	—	—	1
Others	45	—	—	10	2	2	0	7	4	3	9	7
Total	319	2	5	49	34	17	16	36	24	15	62	59

*Recipients identify themselves as in home economics, but degree usually granted by chemistry departments.

Source: First 3 editions of *American Men of Science*.

Another psychologist, Margaret Washburn, was an outstanding graduate student of James McKeen Cattell at Columbia in the 1890s, but since there were no fellowships there for women, she had to transfer to Cornell, where she earned her degree in 1894. Likewise Ida Hyde, a Bryn Mawr graduate, had great difficulty being admitted to the German universities in 1893, but she persevered and became one of the first women to earn a German Ph. D. in 1896.[15] By the late 1890s such incidents became less and less common as most of the major graduate schools liberalized their policies on women and began granting them degrees in large numbers.[16]

Figure 3. Cornelia Clapp (right), a marine embryologist, and some of her students at Mount Holyoke College, about 1893. Miss Clapp was also affiliated with the Woods Hole Marine Biological Laboratory. (Courtesy, Williston Memorial Library, Mount Holyoke College.)

One of the main reasons for a change in the attitude toward women at graduate schools, and the dramatic increase in higher degrees awarded them in the mid-1890s, was the example of the University of Chicago, which was founded in 1892 and from the start accepted men and women on an equal status in both its graduate and undergraduate programs. It rapidly out-stripped the eastern universities in the number of degrees it granted women, as is shown in Table 6. In botany, in particular, it established an outstanding tradition, granting as many degrees in this one department before 1920 as did the next seven universities combined. Chicago even usurped some of what might have been considered Harvard's domain. In its first year Chicago attracted students from 95 institutions, with the largest group--14 students--coming from Wellesley College. [17]

By 1900 women's education had come a long way from the days of Matthew Vassar's pioneering venture. It had largely fulfilled the hopes of those who sought an education for women equal to that of men. A woman who wished to study science in 1900 could find both undergraduate and graduate training of a high order. Her chances for eminence and recognition in her chosen field, however, remained much more limited than was true for men.

Recognition and Professionalism

One of James McKeen Cattell's purposes in compiling his massive directories of American scientists had been to identify a sample of the top 1,000 scientists for further study. For this purpose he had scientists in each field rank each other, and he starred the top 1,000 in his volumes. It was easier to be included in this esteemed group in 1906 than in 1921, but in both years the women's showing was disappointingly poor. Thirty of the 504 women scientists were starred in one of the first three editions--less than half the 73 starred in the male sample. Because their numbers remained roughly constant (19 in 1906, 18 in 1910, and 24 in 1921) at a time when the number of women in the directories almost tripled, the percentage of women who were starred dropped from 12.8 per cent in 1906 to 5.2 per cent in 1921.

Why this drop should have occurred when the group as a whole was undoubtedly highly talented and motivated, and when their numbers and opportunities were increasing so greatly, is not at all clear. Cattell could not understand the phenomenon and thought there must be some "innate sexual disqualification," although he added, "It is possible that the lack of encouragement and sympathy is greater than appears on the surface."[18] Today we would attribute this sexual difference in achievement to discrimination in various forms. The women were employed at women's colleges and in low echelons elsewhere, where they were less visible, had less access to research facilities, fewer stimulating colleagues, and heavier teaching loads. Perhaps worst of all, there was much less incentive for them to persevere to overcome their obstacles, for there was no place else for them to go.

The generally low level of accomplishment among women scientists and their virtual segregation raises the question of whether they really were a part of the profession. Numerically, they were a tiny minority--less than 10 per cent--in most fields except psychology, where they constituted 22.8 per cent of the field, and of course, home economics (see Table 7).

A few prominent women seem to have been readily accepted in their fields, as shown by their election to the major professional societies. Maria Mitchell again led the way when she was elected to the American Academy of Arts and Sciences in 1848 (though opposed by Asa Gray) and to the American Association for the Advancement of Science in 1850 (supported by Louis Agassiz). The California Academy of Sciences went so far as to invite women to join in 1853, the year it was established. Rachel Bodley, Dean of the Women's Medical College in Phila-

Table 7. Percentages of women in various sciences, 1906–1921.

Field	%
Engineering	0
Agricultural sciences	0
Pharmacy	0
Chemistry	2.5
Physics	2.6
Geology	3.8
Medical sciences	4.3
Mathematics	6.3
Astronomy	6.4
Anthropology	7.7
Zoology	8.9
Botany	9.4
Psychology	22.8
Home economics	100.0

Source: First three editions of *American Men of Science.*

delphia, was elected to the American Chemical Society in 1876, the year it was founded, and Charlotte Scott helped establish the American Mathematical Society in 1894. Similarly, Marcia Keith of Mount Holyoke and Isabelle Stone of Vassar were among the founders of the American Physical Society,[20] and Mary E. Pennington joined the American Society of Biological Chemists at its second annual meeting in 1907.

But other women encountered opposition. In physiology there was a fifteen-year lag between the formation of the American Physiological Society in 1887 and the election of its first woman member, Ida Hyde of the University of Kansas, in 1902. Women anthropologists were deliberately excluded at least once by their brethren, but their problems may have stemmed as much from the insecure pre-professional status of their field as from discrimination. In 1885 ten women led by Matilda Coxe Stevenson of the Bureau of American Ethnology met in Washington to form the Women's Anthropological Society of America. Angry at being excluded from the men's Anthropological Society of Washington because of their sex and lack of training, they banded together to present papers to each other as a means of mutual encouragement. They also devised a feminist rationale for their inclusion in the field, arguing that women could contribute to anthropology in ways that men could not, since tribeswomen would theoretically be more willing to discuss certain aspects of their lives with them rather than with men. Such a distinction turned out not to be valid in practice, since, as Nancy Lurie has pointed out, the idea was more in the Victorian anthropologists' minds than in the Indians'. However, the women's group had about 40 active members in 1889 and maintained a schedule of biweekly meetings until 1899, when it quietly merged with the men's group. By then feelings on both sides had softened, and the men were finally willing to accept their female colleagues into the field. In 1903 the combined group elected Alice Fletcher of the Peabody Museum at Harvard, an outstanding ethnologist and advocate of Indian rights, to be its first woman president.[21]

But if the professional societies were generally willing to admit women, the nation's most prestigious group, the National Academy of Sciences, was not. It was not until 1925 that the first woman, Florence Sabin (Fig. 5) was selected and 1931 when the second, Margaret Washburn, was added. These women had done outstanding work (see below), but it can only be considered an outrage that others such as Annie Jump Cannon and Charlotte Scott were never elected.

But being elected to membership in the professional societies is really no test of how willingly and actively women

Figure 4 (opposite). Alice C. Fletcher, an ethnologist, and colleagues on a visit to the Nez Percé Indians in 1890 in order to assign land to them. (Courtesy, The Jane Gay Dodge Collection, The Schlesinger Library, Radcliffe College.)

Figure 5.   Florence Sabin, an anatomist.   She was the first woman in the National Academy of Scientists.   (Courtesy of the American Philosophical Society.)

were accepted into the profession.   As Martha White has pointed out, there is much more to being a member of the profession and feeling professionally involved than merely being elected and paying the dues.   She considered important such factors as equal expectations, equal access to prestigious journals, equal support from eminent sponsors, equally stimulating colleagues, and meaningful communication. 22 Without examining the correspondence of a number of these women, it is difficult to imagine what they felt about these issues.   Some, such as the highly regarded Margaret Washburn, resigned themselves to their positions in the women's colleges, although in private they admitted they would have much preferred coeducational institutions. 23   Others, such as Maria Mitchell, Ida Hyde, and ichthyologist Rosa Smith Eigenmann, spoke out whenever possible against their unequal conditions and the male condescension that repeatedly considered them "good for a woman."

Some of these women scientists even went further and tried to correct some of the unequal conditions facing them. Marion Talbot and Ellen Richards founded the Association of

Collegiate Alumnae in Boston in 1881 (later the American Association of University Women). From its early days the association was concerned with opportunities for graduate work for women. Even though many graduate schools began to admit women in the 1890s, the almost total lack of fellowships for women continued to be a real problem. Several eminent women scientists--among them Christine Ladd-Franklin, Ida Hyde, Margaret Maltby, and Ellen Richards--were active in the association, and they worked hard to establish fellowship funds for women.

By 1920 at least 26 women scientists had benefitted greatly from these awards. For most of them a career in science would not have been possible without the AAUW fellowship. For a number of years, the AAUW women scientists also raised money to support a table for a woman at the Naples Zoological Station, to grant a prize for the best thesis by a woman and, in 1921, to purchase one gram of radium for Madame Curie to help her continue her researches. All these projects were part of their program to gain greater opportunity and recognition for women scientists.[24] Although these women were divided among the several sciences, they faced certain common problems, which they hoped the AAUW would help alleviate for future women scientists.

Achievements

Despite all these obstacles, a few women scientists did important scientific work before 1920. The number of "starred" women varied among the sciences from one each in physics and anthropology to six in psychology and eight in zoology, suggesting perhaps that where the women congregated they achieved the most, although the samples are too small to be reliable. Among the zoologists Mary Jane Rathbun classified crustacea at the U.S. National Museum for over thirty years, and Cornelia Clapp, who worked in marine embryology, was a mainstay of the Marine Biological Laboratory at Woods Hole for a number of summers. A younger group, some of them students of Thomas Hunt Morgan at Bryn Mawr from 1891 to 1904, worked in the area of genetics. Charles Rosenberg has shown that, of the 52 American first authors cited in the bibliography of Morgan's The Mechanism of Mendelian Heredity (1915), seven (or 13.5 per cent) were women (Alice M. Boring, E. Eleanor Carothers, Katherine Foot, Mildred Hoge Richards, Helen Dean King, Margaret Morris Hoskins, and Nettie M. Stevens), and two others (Ella C.

Strobell and Rhoda Erdmann) were second authors.[25]   The
most eminent of this group was Nettie Stevens, who discovered independently of E. B. Wilson in 1905 the relation
of the X and Y chromosomes to sex determination.

In psychology Margaret Washburn of Vassar College
was well known for her experimental work in space percep-
tion and animal psychology.   Mary Calkins studied dreams
and association and developed her own introspectionist "self-
psychology."   Christine Ladd-Franklin combined her interest
in mathematics and psychology in studies of color sensation
and optical illusions.   Helen Thompson Woolley of Cincinnati
pioneered in the study of child development and was a leader
in the vocational guidance movement.   Leta Stetter Holling-
worth, in 1914 the first Civil Service psychologist in New
York City, later moved to Teachers College, Columbia University, where her mental and motor tests of women helped
undermine the view of her colleague Edward L. Thorndike
that men were more capable than women.

In anatomy Florence Sabin, a student of Franklin P.
Mall and the first female full professor at Johns Hopkins
University, made intricate studies of the brain and the lym-
phatic and vascular systems.   In mathematics Charlotte
Scott and Anna Johnson Pell, both of Bryn Mawr College,
worked on algebraic curves and integral equations.   In
physics Margaret Maltby of Barnard College studied the
physical chemistry of electrolytic resistances and dilute solutions and did some work on radioactivity.

Florence Bascom of Bryn Mawr worked on the petrog-
raphy of Massachusetts and eastern Pennsylvania, and Ermine
Cowles Case of the University of Michigan studied the oste-
ology of extinct reptiles, meeting her death in South Africa
in 1923 while collecting specimens.   Ellen Semple was the
author of several books and one of the founders of the science
of geography, although her only professional employment was
as an occasional lecturer at the University of Chicago.   In
botany Alice Eastwood classified thousands of plants as curator at the California Academy of Sciences for almost sixty
years (1892-1949), Elizabeth Knight Britton assisted her
husband at the N.Y. Botanical Gardens in the classification
of mosses, and Margaret C. Ferguson of Wellesley College
studied the fertilization and germination of spores of fungi.

Probably the most famous of all these women were the
astronomers Annie Jump Cannon and Williamina P. Fleming,

both "Curators of the Photographic Plates" at the Harvard College Observatory, where they did outstanding work on the classification of stellar spectra. They also directed teams of up to fifteen other women in this painstaking and poorly paid technical work, which few men would undertake, but which, in the end, did so much to make the Observatory and its Draper Star Catalogue known throughout the world.[26] Surely when women were given the chance, they could perform important work.

Recent work in the sociology of science has shown that there are certain barriers ("stratification") within the institution of science which distort the even distribution of honors and prevent some men from achieving the rewards their ability would seem to merit.[27] For women scientists sexual discrimination was an additional and particularly oppressive form of segregational stratification which suppressed their achievement by at least half and which certainly violated what we all like to think of as the norms of the scientific profession. Unfortunately, we know too little of what drove these women to enter science and to try to achieve in it despite the obstacles. We are only beginning to uncover their side of the history of science.

Notes

1.   These 504 hardly made a complete list, but it is the most convenient one available for both size of sample and completeness of biographical data. The only other possible source of data would have been the Ph. D. lists of the major universities, but biographical data on these persons are more difficult to obtain, and many scientists did not have doctorates. A comparison of the Yale and University of Chicago lists with AMS directories reveals a 50-60 per cent overlap. Cattell obtained his lists from other biographical directories, membership lists of 50 scientific societies, rosters of 70 institutions of learning, contributors to Science, and an advertisement in Science (16:746-47, 1902). These lists may have been sex-biased, reflecting the discrimination of some societies and institutions against women, but Cattell cast such a wide net that he may have overcome this difficulty.

2.   Every eighteenth male yielded a list of 506 names, of

which four were eliminated because they were in nonscience fields--medical illustration, ordnance, physical education, and philosophy.

3. Nine women in nonscience fields--photography, physical education, education, history, and philosophy--were excluded from the tabulations.

4. Alice Hamilton. 1943. Exploring the Dangerous Trades: The Autobiography of Alice Hamilton, M.D. Boston: Little, Brown and Co., p. 36.

5. Yale University. 1961. Doctors of Philosophy, 1861-1960. New Haven, pp. 258-62.

6. The best short summary I have found of the home economics movement is in Alfred C. True (1929), A History of Agricultural Education in the U.S., 1785-1925. Washington, D.C.: U.S. Gov't. Printing Office, pp. 267-72.
   See also Jesse Bernard. 1964. Academic Women. Cleveland: World Pub. Co., pp. 8-11, 97-99; and Caroline Hunt. 1918. The Life of Ellen H. Richards. Boston: Whitcomb and Barrows.

7. Alice S. Rossi. 1965. "Women in science: Why so few?" Science 148:1197 (28 May).

8. "Preliminary report of Committee W, on status of women in college and university faculties." Bulletin of AAUP 7:21-32, 1921.

9. Edna Yost. 1943. American Women of Science. Philadelphia: Frederick A. Stokes Co. pp. 80-98.

10. Fred W. Powell. 1927. The Bureau of Plant Industry: Its History, Activities and Organization. Baltimore: Johns Hopkins Press, pp. 44-76. This volume has extensive salary data, unfortunately for 1926.

11. Dean R. Brimhall. 1922, 1923. "Family resemblances among American men of science." American Naturalist 56:546-47; 57:342-43.

12. Nancy O. Lurie. 1966. "Women in early anthropology." In Pioneers of American Anthropology, June Helm [MacNeish], ed. Seattle: University of Washington Press, pp. 29-81.

13. Helen Wright. 1949. Sweeper in the Sky: The Life of Maria Mitchell, First Woman Astronomer in America. New York: The Macmillan Co. She also had to fight hard to get a salary equal to that of her less distinguished male colleagues.

14. When offered a Radcliffe Ph.D. in 1902, Professor Calkins refused, insisting on a Harvard degree or none at all. See Margaret Munsterberg, 1922. Hugo Munsterberg, His Life and Work. New York: D. Appleton and Co., p. 76.

15. Ida H. Hyde. 1938. "Before women were human beings..." Journal of the American Association of University Women 31:226-36. Margaret Maltby earned a Ph.D. in physics at Göttingen in 1895.

16. See Walter Crosby Eells. 1956. "Earned doctorates for women in the nineteenth century." Bulletin of the American Association of University Professors 42:644-51. Yale's President attributed the decision to admit women to its graduate school to "considerations of fairness, rather than to any widespread demand or far-reaching policy" and not to the example of the University of Chicago. See also Arthur T. Hadley. 1892. "The admission of women as graduate students at Yale." Educational Review 3:486. The author thanks Thomas Cadwallader for the former reference and G. W. Pierson and R. P. von Oyen for the latter one.

17. Richard J. Storr. 1966. Harper's University: The Beginnings. Chicago: University of Chicago Press, p. 109.

18. James McKeen Cattell. 1910. "A further statistical study of American men of science." American Men of Science. 2nd ed. New York: The Science Press, p. 584.

19. The preponderance of women in home economics may have depressed the status of the field as a whole, as Dee Garrison has recently suggested happened in librarianship. See her article "The tender technicians: The feminization of public librarianship, 1876-1905." Journal of Social History 6:131-59 (1972-73).

20. K. K. Darrow. 1949. "The names of those who met on 20 May 1899 to organize the American Physical Society." APS Bulletin 24(5):34 (16 June). I thank Joan Warnow and Charles Weiner of the Center for the History of Physics, American Institute of Physics, New York City, for this and other useful references.

21. See Ref. 12 and Organization and Historical Sketch of the Women's Anthropological Society of America. Washington, D.C.: The Society, 1899; and D. S. Lamb. 1906. "The story of the Anthropological Society of Washington." American Anthropologist 8:564-79.

22. Martha S. White. 1970. "Psychological and social barriers to women in science." Science 170:413-16 (23 Oct.).

23. E. G. Boring. 1971. Margaret Floy Washburn. Notable American Women. Cambridge: Harvard University Press. The NAW has been an indispensable reference throughout.

24. Marion Talbot and Lois K. M. Rosenberry. 1931. The History of the American Association of University Women, 1881-1931. Boston: Houghton Mifflin Co.

25. Charles E. Rosenberg, 1967. "Factors in the development of genetics in the United States: Some suggestions." J. Hist. Med. 22:43-44.

26. Bessie Z. Jones and Lyle G. Boyd. 1971. "A field for women." In The Harvard College Observatory: The First Four Directorships, 1839-1919. Cambridge: Harvard University Press, Chapt. 11.

27. The best summary is Harriet Zuckerman. 1970. "Stratification in American science." Sociological Inquiry 40:235-57.

3. JULIA B. HALL AND ALUMINUM*

by Martha Moore Trescott

I

The Aluminum Company of America (ALCOA) is one of the largest and best-known enterprises in the world. Many Americans are familiar with the story of ALCOA's beginnings and with the name of Charles Martin Hall, the chemist and inventor whose 1886 process for the electrolytic production of aluminum solved the decades-old search for a way to manufacture aluminum metal cheaply.[1] Much less well-known then and now is the degree to which the Hall innovations and the very establishment of the ALCOA forerunner, the Pittsburgh Reduction Company, can be attributed to Julia B. Hall. She was an older sister of Charles and, like Charles, had college training in chemistry. She would have been unremembered and virtually unrecorded had it not been for the depiction of her life and contributions as an adjunct to Charles by Junius D. Edwards in The Immortal Woodshed and for the brief mention given her by Charles C. Carr, company historian of ALCOA, in ALCOA, An American Enterprise.[2] Whereas Edwards and Carr highlighted the genius of Charles, a brilliance from which this paper does not intend to detract, the focus here will be on Julia. This essay suggests that successful invention and innovation have been a team effort, both historically and today, requiring many skills and different kinds of contributions. And in this effort,

*The author is indebted to Professors Paul Uselding of the University of Illinois and Paul B. Trescott of Southern Illinois University for critique of various drafts of this essay. Also, the author wishes to thank the National Science Foundation and the Lincoln Educational Foundation for support of her dissertation project on The Rise of the American Electrochemicals Industry, 1880-1910: Studies in the American Technological Environment, from which the work on the Halls stemmed.

female relatives of inventors, such as Julia Hall, may have played a larger role than has previously been acknowledged. If this is true, the social costs of invention and innovation in the nineteenth century and at other times when invention took place in the home have not been truly understood. [3]

Much of the biographical information on the Halls can be obtained from the works of Carr and Edwards, both of whom knew of and drew upon the letters from Charles to Julia. This essay is a reconsideration of those letters and other primary sources in order to reassess Julia's role in the Hall invention and formation of the Pittsburgh Reduction Company. Her contributions to the establishment of priority of the 1886 Hall invention for the electrolysis of alumina in a cryolite bath over the claims of Frenchman Paul Hèroult will be especially emphasized. While Carr and Edwards have noted Julia's role in brief, implying that her contributions were at best secondary to Charles's (with Edwards' depiction of her as quite inane at times), she in fact played a central role in her brother's projects. It might not be an exaggeration to say that she actually managed many aspects of the invention process which culminated in the foundation of the Pittsburgh Reduction Company.

II

Julia Brainerd Hall was born November 11, 1859, in Jamaica, British West Indies, daughter of a Congregational minister. Her middle name reflected her place of birth, Brainerd Station. In 1860, shortly after her birth, the family, who had gone to the West Indies from Ohio, returned to Ohio because the coming of the Civil War made it necessary for the American Missionary Association to close its foreign missions. Heman Bassett Hall, Julia's father, assumed the pulpit of the Congregational Church in Thompson, Ohio, and in this locale Charles was born December 6, 1863. Julia and Charles had three older siblings as they grew up-- a brother and two sisters--in addition to two younger sisters, and all of the Hall children graduated from Oberlin. [4]

Heman Hall and his wife, Sophronia Brooks Hall, returned to the town of Oberlin in 1873. [5] Sophronia, mother of all of the Hall children and a former Oberlin student, took a very active role in encouraging the higher education of each child. Thus Oberlin, a pioneer in college coeducation in the

U.S., became the alma mater of not only the male Halls, including Heman, but also Julia and her sisters. It should be noted, however, that Julia and her mother and sisters received an Oberlin education which was somewhat differentiated from the courses taken by men. While Junius Edwards notes that Julia took the "classic course," the Oberlin files indicate that Ellen Julia Hall (graduated, 1875), Emily Brooks Hall (1881), Julia Brainerd Hall (1881), Edith May Hall (1889), and Louie Alice Hall (1892) all were enrolled in the "Literary Course," the 1875-1894 extension of the earlier (1836-74) "Ladies' Course." The Literary Course differed from the A.B. program in that the latter "contained a prescribed amount of both Latin and Greek." The graduates of the Literary Course received diplomas but not degrees.[6]

Thus, even though Julia did not receive the undergraduate degree which Charles was awarded, the records show that their courses of study were quite similar, with Julia having completed slightly more credits in science overall at Oberlin than Charles. The only indication of a basic difference in the curricula of Julia and Charles seems to have been the "rhetoric exam," which Charles took at the end of each semester.[7] Apparently, Julia and Charles had some of the same teachers, if the fact that they both took chemistry under Professor Franklin F. Jewett, both in their junior year, four years apart in time, is any indication.[8] Charles had two semesters of chemistry to Julia's one. But, according to their college transcripts from the Oberlin archives, they both completed 38 different "solid" courses. Both had taken one semester each of algebra, trigonometry, astronomy, economics, physiology, geology, and a course termed "mechanics."[9]

In science, therefore, Julia and Charles had had very similar collegiate training. But social expectations were quite different for the two. While Charles went on to use his education to become a "rich inventor," as indeed he intended to do from his undergraduate days, Julia, upon graduation from Oberlin in 1881, assumed the tasks of housekeeping and raising her two younger sisters in her mother's place. Sophronia, who had also attended Oberlin during 1844-46 and 1849-50, had become critically ill by the early 1880s and died on May 7, 1885. It is ironic and tragic that she died almost exactly two months before Charles, whose intellect and education she had so nurtured, graduated from Oberlin. While her illness and death were undoubtedly

a source of sorrow to Charles, they did not significantly
distract him from his search for cheap aluminum.   Julia,
however, became Sophronia's replacement in the household,
fulfilling many of Oberlin's apparent expectations about its
women students at that time. [10]

Julia and Charles had been very close when they were
growing up.   They were continual playmates when children.
And, as Julia was the older, it was perhaps natural for
Charles to be dependent upon Julia emotionally.   This de-
pendency seemed to have functioned throughout his life, if
one examines the letters from Charles to Julia, 1882-1909.
Evidently, she was also very involved with his activities,
as attested by the sheer volume of her apparent correspondence to him--sometimes three or more letters per week.[11]
He usually attempted to write her at least once a week before 1890 when he was away from Oberlin, engaged in one
remunerative pursuit or another (after 1886 his travels often
had to do with his aluminum invention).   Their correspondence continued on a fairly regular basis over approximately
twenty-seven years, although the number and length (and
emotional intensity) of his letters to her tapered off after
the early nineties.   Julia, however, continued to write
Charles often after the early 1890s, evidently sometimes
still several letters per week.

The historians Carr and Edwards did utilize the letters in writing their books, but they by no means included
all of the correspondence.   This reevaluation is based in
part on materials which they omitted or only partially published.   One particularly serious omission consists of an 1887
eye-witness account, "History of C. M. Hall's Aluminum Invention," composed by Julia, which formed the core of her
testimony in the patent interference case between Hall and
Hèroult in 1887.   Her testimony, in fact, assured Hall's
victory in this case, as will be shown later.   Suffice it to
say here that Edwards had this 1887 account typed, along
with the letters from Charles to Julia, and transmitted them
to company historian Carr, but neither historian referred to
Julia's account as such.   To verify her account, this essay
will draw upon copies of testimonies in the interference
case. [12]

Julia's letters to Charles, more numerous than his
to her, have been lost.   She herself instructed him at times
to burn her letters and not to leave them lying around, as
they apparently contained information relevant to the inven-

tions.[13] She was extremely cautious with the letters he sent her, even to the point of censoring important names, dates and other facts in case the letters fell into the wrong hands. But she was careful not to censor technical details which might be needed in court, as will be shown. In fact, this paper maintains that the letters, rather than a random and idle interchange between family members, represent in this case documents whose authors realized at the time of composition and transmittal that they would very likely one day be needed in litigation.

From the letters and other documents, Julia's general managerial role in many aspects of the invention process described here can only be inferred. However, it is clear from surviving records that she did manage much of the information systems surrounding Charles' inventions. In other words, Charles was not a "lone inventor." Rather, Charles and Julia worked together to record and protect the secrecy of Charles' inventive efforts, to find financing for his inventions, and in other ways pertinent to successful invention and innovation. The story of their working relationship, then, needs to be constructed, drawing upon the letters, the court documents and other material as supplements to the information presented by Edwards and Carr.

III

Charles conducted many of his experiments in a woodshed in back of the Hall residence, with the bulk of his work there occurring during 1882-6. With Julia in charge of the household, Charles looked to her for permission to enlarge the area for his work. After Charles's graduation from Oberlin in 1885, Edwards writes, "It was not difficult to persuade Julia to let him enlarge his working space. Washing machine, sawbuck and woodpile only had to be moved back a bit. Most of the woodshed floor was at kitchen level.... Julia's headquarters were in the adjoining kitchen and she had formed the habit of looking in from time to time to see what her brother was doing in his woodshed laboratory."[14] Although there was a great difference in the neighboring "headquarters" of Charles and Julia, it is little wonder that her scientific curiosity, honed by an Oberlin education, should have prompted her to his lab. In fact, the letters indicate that her visits were more than casual and that, very likely, she was present daily in his laboratory, at least for a period during 1885-6 when his experiments were building toward his important discovery in February, 1886.[15]

Whereas Julia's assistance could be viewed as invaluable for the progress of the work, Edwards typically depicts Julia quite chauvinistically as a total adjunct of Charles, taking all her cues from her brother. Close review of the letters does not bear out Edwards' interpretation, but one does have to look closely. Charles tended to claim credit for his ideas, giving little acknowledgment of Julia's help.

Yet the fact that Charles wrote primarily to Julia of his business and technical insights and not often to other family members, including his father, in such a confidential way, suggests a strong dependence upon her scientific and business judgment. In one letter of 1887, for instance, Charles appealed to Julia: "Do you have any advice??" concerning his quandary over whether he should sell his process to Cowles Electric Smelting or to the chemical firm of Grasselli, for how much, and whether he should enter into the employ of Cowles.[16] Charles punctuated the question with double question marks, pressing down on his pen to darken the sentence. Junius Edwards, in typing a copy of the original letter, underlined it.[17] Charles clearly did seek Julia's insights in such decision-making.

On July 27, 1882, Charles wrote Julia during his travels selling copies of The Golden Censer. At that early date, he was not only working on the aluminum problem but also on windmill improvements. He was, however, undoubtedly referring to his aluminum project when he commented,

> I really think the idea quite promising.... When I get over into Hancock County I probably shall not get to the P.O. more than once in two or three days, so you will have to take that into account and keep me informed somewhat ahead, or rather tell me what indications there are.[18]

In this, as in many of the letters, names and other data have been crossed out. Julia was very circumspect not to let even other family members have access to much information concerning Charles's business and technical affairs, and she apparently regularly admonished Charles to be careful about the letters she sent him.[19]

Also in the July 27, 1882 letter, there appears a notation by Charles on the first page which, again, clearly shows her role as adviser and information transfer agent in his business and technical matters.

> I wrote ____ in order to get his influence with
> ____ . Now what I want of you is, shall I write
> about these two things, or none or one to the Sci.
> [Scientific] American. Write me soon.... That
> about Dr. Herrick is true I think. I well remem-
> ber ____ . Keep this letter and all others. They
> may be useful in a future lawsuit or something. [20]

The blank spaces above represent examples of Julia's censor-
ship where she felt Charles's interests might be damaged if
the wrong parties came across these names. Earlier in the
letter Charles had referred to the Scientific American, a
journal very important to inventors at that time. He had
indicated that an inventor could find out whether a particular
invention was new "by writing to the publishers of the Scientific American and asking for advice which they give free.
They are safe patent agents having been in the business for
40 years. "[21] Dr. Herrick appeared to be a family friend
and was probably a relative of the Mr. Herrick to whom
Charles repeatedly referred in 1887 as being in the chemical
business and a potential backer for the aluminum process.

Julia was not only an adviser and confessor to Charles
on personal, business and patent matters; he also discussed
the technology and science behind potential inventions with
her in great detail. Thus in the July 27, 1882 letter, he
also commented:

> ... [T]he tubes would clog the fire to which I say
> they could be put around the fire perhaps and
> gather the waste heat. Another could get suffi-
> cient air to them to which; [sic] make them large
> enough. The tall chimney creates the draft, the
> air will enter through the easiest passages, regu-
> late the dampers and make the air enough enter
> through the tubes. [22]

It is not clear from the text of the letter to what he is referring, since Julia had crossed out several lines. However,
Charles was integrally interested in furnace and heating innovations later in the electrolysis of alumina in a molten
bath. For instance, on December 18, 1888, Charles ex-
pounded in similar detail to Julia on technical problems of
heating the aluminum reduction pots. By this time, the
Pittsburgh Reduction Company had been organized and operating for several months (July 31, 1888 was the date of its
establishment), and Charles was reporting on certain early

difficulties with the pots. His method of external heating of the pots was causing leakage, and this letter documents one of the most significant transitions in the early days of Pittsburgh Reduction Company--from external to internal heating, i.e., reliance on the heat of the electric current itself by employing thicker linings for the pots. The large improvements in efficiency resulting from this change truly astounded Hall and his co-workers, and he was able to discuss the technicalities involved with Julia, as if she had been on the job in the plant with him.[23]   This predates by several years his involved technical correspondence with Alfred Hunt, the entrepreneur who helped organize Pittsburgh Reduction Company, so Julia was still at this time apparently serving as a scientifically knowledgeable correspondent and possible adviser.

It is certain that she was still a source of help with patent matters at this time and later, for in 1891 he wrote:

> This letter I think you better preserve and also the enclosed wires or tubes with it. You will notice that these wires are really tubes with an aluminium core. It is only necessary to soak them a few days or perhaps less in Hydrochloric [sic] or Hydrofluoric [sic] acid to remove the aluminium and leave them hollow all the way through....
>
> I have tubes now enough to make a battery, that is, a small one. They are about of the right composition, that is, the old alloy that you know about....
>
> Please let me know if you receive this letter and also the two short tubes which I enclose.[24]

Charles continued until he died to be preoccupied with construction of a battery, a kind of fuel cell which would serve as a source of cheap electricity for his aluminum process and other uses.[25]

The letters, ranging over almost a decade, 1882-1891, thus indicate not only a brotherly concern for a sister, but also--and perhaps primarily--they were often intended to be records for potential patent litigation, a phenomenon which Hall and his company more than once experienced. Also, the 1887 eye-witness account written by Julia concerning the Hall aluminum invention and her testimony before the patent examiner in October, 1887 tend to corroborate this interpretation of the letters.

Because of the importance of these legal documents
and because they have not been published before, they will
be quoted at length. They not only document Julia's role
but also detail the process of invention as well as many of
the technical aspects. They demonstrate precisely how
Julia's education and her familiarity with chemical matters,
knowledge acquired at Oberlin, served Charles. And her
eye-witness testimony served to secure Hall's victory in the
interference case with Hèroult, as even Edwards admits.[26]
In her written history of the invention, Julia says:

> I went with my brother to the shed where he had
> a furnace. He had a quantity of cryolite in a
> crucible. I noticed that it was clear and dissolved
> alumina readily as he put it in.
> That same day, February 10, 1886, Charles
> commenced to prepare apparatus and materials to
> test the value of his discovery. I helped some in
> washing alumina, etc. By Tuesday of the next
> week his materials were ready and he conducted
> his first experiment.
> ... I saw the experiment in operation and
> also saw the contents of the crucible after it was
> poured out. My brother did not find that he had
> made any aluminium. [Julia, Charles and contem-
> poraries typically spelled the metal that way.]
> Every day after that for some days he performed
> experiments, all of which I witnessed as on the
> first day....
> On Tuesday, Feb. 23rd, 1886, a week from the
> day on which he made his first experiment, my
> brother Charles conducted a successful experiment
> by which he produced aluminium....
> I witnessed the experiment, saw the mixture
> after it was poured out and, after it was cooled,
> picked out from it a number of small globules of
> aluminium.
> I knew it was aluminium from its bright
> silvery color and light weight and from the fact
> that my brother was able to dissolve the globules
> in hydrochloric acid....
> In the evening of the same day Charles wrote
> a letter to my oldest brother, Rev. George E. Hall
> of Dover, N.H. in which he told him of his inven-
> tion and described to him what it was and told him
> of the results of his experiments.
> I read the letter before it was mailed....

> On the 24th of Feb. 1886, Charles conducted
> further successful experiments, and in the evening
> again wrote to G. E. Hall. I <u>read</u> the letter be-
> fore it was mailed--it was almost entirely about
> my brother Charles' aluminium invention. . . .
> Nearly all the letters written to [George] I
> <u>read</u> before they were sent and all the letters re-
> ceived in reply from G. E. Hall I <u>read</u> at the time
> they were received. This correspondence had
> reference almost entirely to C. M. Hall's alumi-
> nium invention and to finding some person to fur-
> nish money for large experiments and patents.
> On Thursday, April 15, 1886, I copied a
> paper for my brother. It was mainly a descrip-
> tion of his invention and he intended it to be shown
> to some specialist in the East. It was finally
> shown to Prof. Cook of Harvard College. . . . I did
> not date this paper but stood by my brother and
> saw him date it. He did not date it very clearly.
> <u>I told him he might want to use the paper some
> time</u> and so copied the last page a second time for
> <u>him.</u> I saw my brother Charles date this paper
> April 15, 1886, and sign it. (Italics are this
> author's.)[27]

It is useful to underline the verbs such as "witness," "saw,"
"read," as they tend to indicate that both Charles and Julia
realized at the time the importance of Julia's being an eye-
witness to the experiments and to the recording and documentation of the experiments.

This account formed a part of Julia's testimony in the
case with Hèroult of France. Hèroult's French patent for
virtually the same process (though arrived at quite differently)
had been issued on April 23, 1886. Since Hall had filed his
claim more than a month after Hèroult had filed in the U.S.
--on July 9, 1886; Hèroult filed in May--Julia's testimony
was very important in proving priority of invention for the
Hall interests.[28]

But how important, in fact, was her testimony here?
According to a copy of the testimonies of witnesses in the
interference case, there were five witnesses. The first was
Charles Hall himself, who described his invention and the
events surrounding it in detail. His testimony amounted to
somewhat over four pages of double-spaced type. Julia was
the second witness, and her testimony ran somewhat over

three such pages. Whereas not even Charles himself could positively identify the exact date when he first disclosed fully the details of his invention to another human being--that person being Julia--Julia herself positively identified the date as February 10, 1886. Upon being questioned just how she knew it was the 10th of February, she gave a lengthy and detailed account of the events surrounding the disclosure. And it is well to cite this particular answer below because it shows her keen awareness of the importance of exactitude of dates and events in the record-keeping, when Charles was somewhat careless about such matters. (It also suggests that she was an active, interacting and people-oriented person.)

Thus the following excerpt from the hearing is important to reproduce here for our purposes. After two questions for purposes of identification, Julia was asked, "Did Charles M. Hall ever describe to you an invention that he had made for the production of Metallic Aluminum [our copy of the transcript of the interference case spells the metal this way]; if so, please state, as nearly as you can recall it, when and where he did so."

A. He did describe to me his invention on the morning of the 10th of February, 1886.

Q. You may please give your reasons for stating that it was on the 10th of February, 1886, when your brother described to you his invention referred to.

A. I went to Cleveland on the 5th of February, 1886. I know that it was the 5th of February, because it was just in the middle of what is called the February vacation of the Cleveland Public Schools, which commenced that year on the 29th of January, and lasted for two weeks. The vacation began Friday, and I went the Friday following, which makes the day the 5th of February. I returned the following Tuesday, which was the 9th. It was the next morning the 10th of February that my brother described to me his invention for the manufacture of Aluminum. 29

Thereafter, she was asked several questions whose answers disclose a myriad of technical details, and these statements would serve as an excellent source for historians interested in the technology here. They follow closely her eye-witness account cited above. The final questions put to Julia concerned her education:

>Q. Are you a graduate of any school or college?

>A. I graduated from the literary department of Oberlin College in 1881.

>Q. Was [sic] Chemistry and Electricity among your studies at Oberlin College?

>A. They were.

>Q. All the time you read the letters referred to, did you understand the subject upon which your brother was writing?

>A. I did. [30]

After Julia's testimony, there were three other witnesses in the following order: their father, Heman B. Hall, Professor Franklin F. Jewett, and another Oberlin professor, Charles H. Churchill. Each of their testimonies ran approximately one-third as long as Julia's, and they variously served to identify Charles and verify the novelty of the invention and to indicate that Charles had worked on it between January and April, 1886. And Heman stated that he "would not attempt to give a critical account of all the steps that he [Charles] took to produce the metal." Charles's brother George did not appear as a witness. [31]

Several additional points should be made about the testimonies, especially Charles's, for the case not only discloses "facts" about the invention and invention process but is also revealing about Charles's personality and nature. Whereas Charles could only say "on or about" February 10, Julia pinpointed it exactly. It may be recalled that February 10 was indeed not the date of the first successful production of aluminum by Charles (February 23, 1886), but to the examiner the date of first full disclosure to another human being was significant. And the fact that she could so clearly designate that date lent credence to the date of reduction to practice. The only other hard evidence Charles would have had of production of aluminum on February 23 centered in the fact that some hours after he had made aluminum, he took a sample to show to Jewett, who had not been an eye-witness of the production itself. Heman Hall claimed to have seen the aluminum globules when they were made, but, of course, not being scientifically knowledgeable, he could not have vouched for their identity. [32]   Another very interesting point about Charles's testimony of 1887 is the fact that he claimed that he did not realize at the time he wrote George Hall that these letters might be used in litigation, [33] as indeed they were.   Julia, however, certainly seemed cognizant of such a contingency, judging from her precision in witnessing Charles's letters as well as his experiments and papers. Recalling the letter of July 27, 1882 (cf. page 155 above), in which Charles explicitly stated that that letter might be useful in a lawsuit, it is highly doubtful that he failed to realize the letters to George might one day serve as legal evidence.   Either he did not tell the truth here, or he was extremely uncomprehending, as Julia stood over him when he wrote and dated the letters to George.

On the contrary, the evidence combines to suggest that Charles and Julia labored together as a team to record the steps in the Hall invention, aware of the importance of secrecy and of possible patent difficulties.   The Halls' discovery of electrolytic aluminum was no bolt from the blue. It was, rather, carefully planned to culminate in a patent or cluster of patents (as actually happened by clever delaying tactics on the part of Hall's attorneys), [34] and ultimately in the establishment of commercially viable production.

IV

Although it is not the purpose of this essay to chart the rise of Pittsburgh Reduction Company, it can be said that after a few initially difficult years, the ALCOA forerunner became a foremost success.   And, of course, Charles Hall realized his dream of becoming a "rich inventor someday."   At the time of his death in 1914, his annual income from ALCOA stock alone amounted to approximately $170,000. [35]   In his will, which provided over $3,000,000 for Oberlin (to which he had already given large sums), most of his wealth was distributed among various educational institutions, along with 200 shares of ALCOA stock to Arthur Vining Davis, also of ALCOA.   (The will named Davis as an executor and trustee of Hall's estate, also.)   But Hall left virtually no money to his relatives, since he had already given them some stock.   "So to his sisters, Edith, Julia and Louie, he gave no money with the following explanation: 'They are already abundantly provided for (through stock) and

any addition to their property would increase their anxieties
and not add to their happiness.'"36

Apparently, Charles did give stock to his family in
1889, in the early stages of the company's operation, but
not typically thereafter. Junius Edwards cites a "block of
four hundred shares" given to Heman and Julia, Edith and
Louie, and adds that "it is not surprising that the largest
certificate was registered in Julia's name."37 As Edwards
indicates, these shares continued to be held by family members after Hall's death. From later records of ALCOA, it
is clear that Julia originally received 100 shares at par
value (which Edwards and Carr state to have been $100 from
1889 into the late nineties, even though in many of those
early years the company was not profitable), exactly 25
shares more than were given to each of Edie, Louie and
Heman.38 Although Julia did indeed receive the largest
amount, it was not greatly in excess of that given the others,
who, by Edwards' own admission,39 did not participate in
Charles's projects to anywhere near the extent that Julia did.
Also, it is interesting that, according to Edwards, Charles
gave brother George 100 shares in 1893 and another 100 in
1894, but George "disposed" of his stock.40 Yet George
and Charles were apparently never close and, in fact, had
a severe falling out when George briefly became involved in
an unsuccessful effort to locate financial backing for Charles's
experiments and patentable process.41

There is no evidence from the records that Julia acquired more shares except through stock splits--one which
doubled her number of shares in 1904 and one which increased
them by five times in 1909. In 1909 the records show that
150 were transferred to each of Edie and Louie from
Charles and 200 to Julia from him.42 However, there is
some indication in the letters from Charles to family members that money was at times being sent to him from the
"girls."43 Therefore, it is unclear as to whether he gave
them these shares in 1909 or whether they bought them. In
any case, the balance of Julia's holdings about the time of
his death was 1,200 shares, as opposed to 900 each for Edie
and Louie. Edie and Louie also each had income from their
jobs as school teachers and, in addition, Edie married.44
So from the standpoint of financial security, the two younger
sisters may have actually fared better than Julia. For the
times, however, Julia was certainly economically well off,
with approximately $7,000 annual income from ALCOA stock
at the time of Charles's death.45

Julia's absolute financial rewards are really not at issue here. Indeed, one can hardly quarrel with Charles's statement in the will that the sisters were "abundantly provided for (through stock)." The point, rather, is the relative rewards received by family members, as those seem to negate the marginal productivity basis of incomes. Indeed, if one can assume that the early stages of Charles's invention and innovation were essentially home-based, then could one not compare this setting to Menlo Park or to later groups in a corporate research and development department, such as at Union Carbide? If so, who then appear as the key figures, engaged in research and development and testing, day in and day out, in the lab? Charles and Julia. In addition, who served as competent and credible eye-witness and recorder of information pertaining to both the experiments and the location of potential backers? Julia. Brother George was involved only for a few months in 1886, with the full knowledge and advice of Julia, and George and Charles never really overcame the rift which resulted from this unsuccessful venture. Heman and the sisters were only peripherally involved. Julia seemed to be the only one, other than Charles, in this minister's family who was scientifically inclined, and certain hard feelings about Charles's flight from religion on the part of Heman are revealed in the letters. Thus, in this large family, Julia was perhaps the only real confidant Charles had.

One might feel that this case is not very representative for a study in women's history because Julia's sex is irrelevant here. A brother could have done the same things, it might be claimed. Yet Charles's brother did not, nor did his father. Also, it may be worth noting that Julia, as a well educated and healthy human resource, was a virtual captive in the home because she was a woman and there were certain expectations concerning her "proper" roles in life. A brother, if as healthy and well educated as Julia, would have been expected to be earning an income. Indeed, as a female, Julia was not only expected to serve in the home but likely praised as well for her devotion to her family, including Charles. Charles, therefore, had at his disposal a devoted, loyal, intelligent and scientifically educated person to assist in his invention. Such a person, if as healthy and educated as Julia, would typically not have been male because a man would not have been in the home in Julia's role. Further, considering the matter of relative rewards, it is doubtful that Julia's later technical and business counterparts in Charles's life and work, such as Alfred Hunt or Arthur V.

Davis, would have settled for as little as Julia did for the tasks she performed. In fact, both Hunt and Davis became well-known and received explicit salaries for their work in the company.

Julia received no similar recognition or pay as direct compensation for her intellectual and managerial contributions. Indeed, she was not even mentioned by Charles in his recollections of the family involvements in his invention and patents during his acceptance of the cherished Perkin Medal in 1911. For that talk, he had been explicitly asked "to give an account of some of the more personal and unpublished facts in connection with my invention of the aluminum process, and of the work of putting it on a commercial basis."[46] In his detailed account, he mentioned his father, Professor Jewett, and his brother, but Julia's name did not appear, not in conjunction with the invention or patent problems nor in reference to his difficult decisions concerning Cowles and Grasselli, all problems in which Charles had relied upon Julia's help and insight.[47] For whatever reasons, he did not credit her help there, in the letters, or in his will.[48]

V

Charles's silence on these matters has undoubtedly been a large part of the explanation for Julia's obscurity. It would, therefore, seem appropriate to consider some aspects of the man's personality in contrast to Julia's. He seems to have been a rather reclusive person, much more so than Julia. This contrast, perhaps more than any other evidence, should make one suspect that while Charles may well have been the major "idea man" as far as the science of the invention was concerned, Julia was probably far more aware of the people involved and the steps which should be taken in managing the secrecy of the invention and recording it minutely for possible litigation later.

Charles was noted for his tendency to isolate himself in the lab, even at Pittsburgh Reduction Company, to isolate himself in his cold apartment at Niagara Falls later on when the company had moved there after 1894, to play his piano alone for hours on end and pour over his Encyclopaedia Britannica, which seemed to be almost a fetish with him. He made a point of not smoking or drinking and would reputedly allow only one man who smoked, Charles F. Chandler, in his residence.   He had a reputation for being a penny-pincher in his daily life, though he gave great sums to Oberlin. [49] One of the few luxuries Charles seemed to allow himself was world travel, and he traveled widely, surely in connection with his business but also for pleasure.   Yet in 1902, Julia noted "his kindness to his sisters" when he wrote a brief note suggesting that the "girls" take a trip to Bermuda.   "If I really thought you would go I might offer to pay your expenses.   It would cost about $90...."[50]

Concerning Charles's personality, two other bits of evidence are instructive.   First, a comparison of his 1911 address as Perkin Medal recipient can be contrasted with the brief but vastly more lively and human remarks of Paul Hèroult, who was also present and whose similar process for producing cheap aluminum had not been upheld in the 1887 interference case with the Halls. [51]   Second, Junius Edwards reproduces several photographs of Charles, taken at various periods of his life, and in these and others seen in additional sources Charles's face is not only extremely youthful and immature-looking but also blank, unlined, and unsmiling. [52]

Julia, on the other hand, seems to have been much more personable.   Not only did she mother Charles and her two younger sisters, she also took care of her father until he died at the age of 88 early in 1911.   The letters from Charles to Julia testify to some extent about her interactions with friends and acquaintances.   Indeed, this essay maintains that she was a virtual center of information for Charles, using her knowledge of others and contacts to best advantage to help him find backers for his inventions, especially in the eighties.   For a shy young man without many connections, whose father and older brother were ministers and not at all scientifically inclined, family and family friends would have been crucial in locating financial backing for the aluminum process.   And indeed this was the case, with Julia's help.

It will be recalled that apparently there was a marked decrease in the number of letters per week Charles wrote Julia after 1890.   One might be tempted to explain this phenomenon by the fact that he often did not feel very well. (He died in 1914 at the age of 51 because of a spleen disease.) But if one examines the volume of his correspondence during the nineties to "my dear Captain" Alfred Hunt, the entrepreneur who helped organize the Pittsburgh Reduction Company, one feels that perhaps Hunt took Julia's place in Charles's attentions to some extent.[53] The analogy of Hunt to Julia, conjectural to be sure, probably can be used to illuminate further the nature of the relationship between Charles and Julia, in that Julia seemed to serve as a kind of entrepreneur, decision-maker and manager, an intelligent and scientifically educated (as was Hunt)[54] sounding board for Charles's ideas. The substitution by Charles of, first, Hunt for Julia as a sharer of business and technical information and, later, of Arthur Vining Davis as he came to work at Pittsburgh Reduction, and finally of Charles F. Chandler, well-known chemist, as Charles Hall grew to fame in the profession at large, may in part explain Charles's apparent change toward Julia.[55] His later letters to her do not contain the kind of technical detail which can be seen in those of the early and mid-eighties. And as early as 1889, shortly after the formation of Pittsburgh Reduction Company, as his days were beginning to be consumed increasingly with company matters, he wrote her, "I am tired of boarding and if I ever get rich am going to have an establishment of my own. Perhaps if you don't get something you like better I will get you to keep house for me..." (Junius Edwards, in citing portions of this letter, entirely omitted the reference to Julia as Charles's potential housekeeper).[56]

Before the formation of Pittsburgh Reduction, when Charles was more dependent upon Julia as a business and scientific assistant, he could temporarily forget her sex and allow her to function as more of an equal. But as his business associates became male, as the company was founded and grew, he began to refer to her as "girl" in the letters and to comment upon her knowledge of social etiquette and her appearance.[57] Perhaps his evidently changed relationship to her by 1911 might help explain his "lapse" of memory in his acceptance speech for the Perkin award.

His silence about her help might also be explained in part in other ways. He may well have taken her for granted, and she may have indulged this attitude. They were both, after all, conditioned to the views of women's roles upheld by such compendia as The Young Lady's Friend and The Golden Censer (which, incidentally, Charles once sold for several months in 1882), that "boys are naturally wiser than you" ("you" meaning young ladies), "sisters should be always willing to attend their brothers, and consider it a privilege to be their companions," "sing merrily while you build the nest," "man [male] is the creature of interest and ambition."[58] And both Charles and Julia had been educated at Oberlin, at a time when the school was imbued with sexist expectations.

Further, at this time there were very few women chemists or other scientists, very few women medical doctors, and even very few female college graduates. And those who did manage to receive undergraduate degrees or the equivalent and who, in some cases, went on to obtain professional and/or graduate degrees were greatly hampered in their pursuit of careers by prejudice and social expectations. [59] Yet it is useful to recall here also that this was a day in which many women were rebelling against the social, economic and legal constraints of their inferior position, as the nineteenth-century women's rights movement snowballed.

Finally, to help explain Charles's failure to recognize overtly Julia's assistance, one can point to the phenomenon of "lost women," that is, those whose contributions were little known because they were somehow attached to a "great man." This has certainly been the fate of engineer Emily Roebling, wife of the famous John Roebling, until very recent years. And many other such "lost women" in various trades and professions could be named. [60] It may be that these women's contributions were "lost" because they were taken for granted or because society was simply conditioned to view the man as the primary achiever, rather than interpreting a given man's greatness as due to the extra advantage he had in the help of an achieving and insightful wife, mother, sister or other female companion. It has been the contention of this paper that, in fact, Charles Hall possessed such an advantage in having Julia's help so close at hand.

One wonders, too, if in the men whose women were "lost" a certain amount of jealousy and resentment did not in fact operate. For in the case of Charles and Julia, there is no evidence in his letters to her that he ever recognized in a complimentary way her scientific and business contributions. Indeed, at times he was quite condescending. It will be recalled that as early as 1889 Charles had offered her the position of housekeeper for him and said that she might "go off on a tangent again" if he failed to write more often. [61] On May 3, 1891, he remarked that "your usual letters came last week.... Also the paper with the article which is in all the Electrical [sic] papers of the country. I have seen it in four. It is not inspired by the Cowles Co., or the Hèroult interests as you will see if you read the

continuation of the same article.   It is only half in the paper
you sent."[62]   Taken in isolation, this might not seem so
condescending, but it is a tone which is evident in many of
the letters from 1882 on, even regarding such matters as to
how to plant strawberries and how to instruct Edie and Louie
to read (despite the fact that by that time Julia was an Oberlin graduate and an educated woman educating women).[63]

Neither Charles nor Julia ever married, it should be
remembered.   Julia was an older sister who served as something of a mother figure upon whom Charles had been emotionally and otherwise dependent.   Julia's style was more
interpersonal than Charles's.   Could he, like his contemporary
in the electrochemicals industry, Edward Acheson, have been
jealous of his female relative's personality and skills?[64]
Indeed, there is evidence that both Margaret Maher Acheson,
wife of Edward (who founded Acheson Industries), and Julia
Hall lived and labored alongside quite humorless and probably
jealous men.   And the women who have, like Julia Hall,
Margaret Acheson, Zula Sperry[65] and others, been in charge
of "building the nest," bearing and raising the children,
catering to the needs of the entire family and attending to
people in general--the women whose primary vocation has
long been in the realm of developing interpersonal systems
which function to guide people through birth, life and death--
have undoubtedly been the objects of considerable misogyny.
The women have been resented by the men, perhaps in part
because the women were often a life-giving, personable and
communicative sex.   Maybe Charles, too, somewhat resented
Julia's person-centered life because his was so devoid of
friends and love.   And perhaps his resentment and jealousy
of her help account for his inability to recognize her assistance.

VI

In many ways slighted or bypassed by previous historians of technology, the female relatives of inventors have
contributed immeasurably to what has ultimately been considered the achievement of a great man or genius.   Fortu-
nately, historians are reviewing and revising earlier, heroic
theories of invention.   Indeed, this reassessment of Hall's
achievements should fit in well with studies of other inventors, such as Edison, which downplay the hero and genius
and seek to integrate the famous into a system of information and interchange in which research and development

proceeds in a planned fashion with an eye toward the market. [66]   Charles and Julia did work together as a team, bent upon defense of priority and secrecy almost from the beginning of serious experimentation.   And, without Julia's presence, Hall very probably would not have been able to establish priority of invention over Hèroult, as Edwards himself admits. [67]

It is time to reassess the social costs of invention and innovation from the standpoint of the women.   It is a truism to cite the high expenditures of our society and other countries today on scientific research and development and to believe that these are substantially higher than during the earlier phases of industrialization.   Yet it may well be that we simply have not adequately accounted for all the human capital which went into the formation of physical capital. And here the efforts and rewards of people such as Julia Hall and other females who contributed to innovation but who have not been adequately compensated and recognized enter in.

A point implied by such reassessment is: do we want more and better technology, such as cheap aluminum products, regardless of the social costs?   Undoubtedly, the world would have had cheap aluminum if Charles Hall had not given the world "his" process, as Hèroult's was already being used abroad.   How much more valuable for our society would it have been had Julia been able to use her skills and education in the marketplace?   The world might well have had two Hall chemists instead of one, with Julia's having lived nearly twelve years longer than Charles, also.   Indeed, the Hall case raises long-standing questions again, such as what should be the opportunities for our professionally inclined women?   Who should take care of the children and family? Why not both men and women?   How should men and women work together, both in the home and in the market?   How will society value technological change vis-à-vis changes in the work environment?

These are very broad questions which this essay can only mention.   Yet it is hoped that this paper will contribute toward a beginning in formulation of further research and possibly some satisfactory answers.   This reassessment of the Hall case seemed particularly important as an aid in understanding the process of invention--planned invention under secrecy--to show that it takes a team and not lone inventors to perform all the tasks necessary to bring an

invention to successful commercial exploitation. And, finally, it has been important to suggest that in the days when
the home may have been a more significant setting for inven-
tion--before the coming of modern corporate research and
development--the wives, mothers and sisters like Julia Hall
played a much larger role than has previously been acknowledged.

Notes

1. "The Search for Cheap Aluminum" is a major theme
    stressed in the author's dissertation, The Rise
    of the American Electrochemicals Industry,
    1880-1910: Studies in the American Technological
    Environment, chapter three. Also, Joseph W.
    Richards, Aluminium, Its History, Occurrence,
    Properties, Metallurgy and Applications, Including
    Its Alloys, 3rd ed. (Philadelphia, 1896) contains
    a first chapter which gives a good bit of detail
    about this search.

2. Junius D. Edwards, The Immortal Woodshed (New York,
    1955) and Charles C. Carr, ALCOA, An American
    Enterprise (New York, 1952).

3. Martha Moore Trescott, "Julia B. Hall and Aluminum,"
    Journal of Chemical Education, 54 (January, 1977),
    24, begins to raise some of these issues, as in the
    oral presentation of this material before the Society
    for the History of Technology, Washington, D.C.,
    October 19, 1975, abstracted in Deborah Shapley,
    "History of American Technology--A Fresh Bicen-
    tennial Look," Science, 190 (November 21, 1975),
    763 and in Technology and Culture, 17 (1976), 507.
    Akin to this notion of viewing social costs of innova-
    tion and technological progress is a conjecture by
    Paul Uselding in the essay on "Manufacturing" in the
    forthcoming Dictionary of American Economic History,
    in which he says that "American textile manufactur-
    ing became price-competitive with British imports"
    only when women and children entered the factories.
    Women's wages in the U.S. and Britain were about
    the same in this period, whereas the wage rate for
    men was higher in the U.S. So, to some degree,
    women's labor enabled the growth of this early
    American industry (p. 24 of typescript). Out of

discussions with Professor Uselding on women in invention, innovation and industry, both conjectures-- that is, the social costs of innovation ideas in the Hall piece and the female textile workers' contribution (albeit indirect) to industrial growth--have emerged.

4. Edwards, The Immortal Woodshed, pp. 4-7 and Carr, ALCOA, pp. 7-12.

5. Heman and Sophronia had been married on November 6, 1849, in Carlisle, Ohio, near Oberlin. She had been born in Carlisle in 1827. Edwards, ibid., pp. 4 and 10.

6. Letter from W. E. Bigglestone, Archivist, Oberlin College Archives, to Martha Moore Trescott, January 20, 1975; also, letter from Miss Gertrude Jacobs, Assisting in Alumni Records Office, Oberlin College, January 23, 1975. For references to Julia's work in chemistry, see Edwards, The Immortal Woodshed, p. 51 and Carr, ALCOA, p. 10.

7. I am indebted to the assistance of Miss Jacobs in obtaining copies of official Oberlin transcripts of both Charles and Julia.

8. Edwards, The Immortal Woodshed, chapters 3 and 4. Cf. also notes 6 and 7 above.

9. See Oberlin transcripts, cited in note 7 above.

10. Edwards, The Immortal Woodshed, pp. 26 and 51, for example, where Edwards cites (p. 51) an 1882 letter from Charles to Julia on the subject of Julia's raising of Edie and Louie, in which Charles instructs Julia (by then an Oberlin graduate) on which books are the "right kind" for girls to read. On the subject of Sophronia's health, see Edwards, pp. 4-7, 46 and also Carr, ALCOA, p. 8. For some information on Sophronia's Oberlin education, cf. letter, Bigglestone to Moore Trescott, cited in note 6 above. Concerning Oberlin education and its effects on the women who attended Oberlin during this period, see Eleanor Flexner, Century of Struggle, The Women's Rights Movement in the United States (New York, 1974), p. 30.

11. Her greater number of letters to him (as compared
     with his to her) can be seen, for example, in letters
     from Charles to Julia dated April 28, 1889, p. 1,
     May 5, 1889, p. 1, May 26, 1889, p. 1, and May
     3, 1891.

12. Julia B. Hall, "History of C. M. Hall's Aluminium In-
     vention," an essay of six pages, written on October
     14, 1887, at Oberlin, Ohio, as Julia's testimony in
     the Hall patent interference case with Hèroult, pp.
     2-3. This account was found among the personal
     papers of Charles M. Hall, including letters to Julia,
     Alfred Hunt and others which had been used by Ed-
     wards in writing The Immortal Woodshed. He had
     had the documents typed. The author is indebted to
     Mrs. Anna G. Lydon, archivist of ALCOA, for help
     in obtaining copies of these materials. She also
     sent the author copies of the interference testimonies,
     taken on October 24, 1887, in the U.S. Patent Office.

13. See, for example, Charles M. Hall to Julia B. Hall,
     letter April 28, 1889, p. 1, ALCOA Archives. In
     this essay, some of the evidence will necessarily be
     circumstantial, and I am indebted to Mike McMahon,
     Director of Historical Programs of the Franklin In-
     stitute, for his suggestions to me following the pre-
     sentation on "Julia Hall and Aluminum" at SHOT
     (see note 3 above) on ways in which the circumstan-
     tial evidence might be buttressed. We can estimate
     the probable number of her letters because of
     Charles's many references in his letters to the
     quantity she wrote him.
          It is perhaps a truism that women's history often
     has to proceed on the basis of viewing circumstantial
     evidence. We can conjecture, for instance, about the
     reasons why Charles's letters were so carefully pre-
     served when hers were not. Julia undoubtedly wished
     to preserve Charles's letters, in part, because he
     had become famous and she had hoped to write a
     history of his inventions. Edwards notes in a typed
     statement in a "'Foreword' to the Letters of Charles
     Martin Hall," written by Edwards July 20, 1936,
     shortly before he had the original Hall letters typed
     (on August 10, 1936), that Julia had desired to write
     a biography of Charles. Louie Hall also assisted
     later in the preservation of the letters and the trans-
     mittal of them to ALCOA, according to Edwards (The

Immortal Woodshed, p. 242). Further, Julia had been conditioned to be a helpmate to her brother and to look up to him, whereas he had not been conditioned to revere her in this way.

Delving into aspects of why he would not have kept her letters brings up certain interesting questions which lead into psychohistory. He may just not have valued her input, taking her for granted. But failure to recognize her contributions may point to an act of will in not preserving her letters to him. (I am indebted to a referee for Technology and Culture for noting the relevance to psychohistory of some of these conjectures.)

14. Edwards, The Immortal Woodshed, p. 51.

15. Julia's statements in "History of C. M. Hall's Aluminium Invention," in conjunction with her oral testimony, leads one to suspect her daily presence, in addition to Edwards' own descriptions, which are somewhat fictionalized, and chauvinistic accounts of the invention process given by Charles and Julia in their testimonies.

16. Charles M. Hall to Julia B. Hall, letter, July 13, 1887, p. 2 of letter.

17. In the typed collection transmitted from Junius D. Edwards to C. C. Carr, August 10, 1936, held now at ALCOA archives (cf. note 12).

18. Charles M. Hall to Julia B. Hall, letter, July 27, 1882, p. 4 of letter.

19. Charles M. Hall to Julia B. Hall, letter, April 28, 1889, p. 1 of letter and the letter dated May 5, 1889, p. 1 of letter, for example.

20. Charles M. Hall to Julia B. Hall, letter, July 27, 1882, p. 1 of original and p. 5 of typed.

21. Ibid., pp. 3-4 typed.

22. Ibid., pp. 5-6 typed.

23. Charles M. Hall to Julia B. Hall, letter, December 18, 1888, pp. 1-2; the date of organization of the

Pittsburgh Reduction Company is cited in note 1. It perhaps should be noted here that this letter may well be the only record of this development, according to Junius Edwards, A Captain in Industry, (New York, 1957), p. 41, where the November 25, 1888 letter is discussed.

24. Charles M. Hall to Julia B. Hall, letter, February 1, 1891, pp. 1-2.

25. See Edwards, The Immortal Woodshed, particularly chapters 11 and 14.

26. Ibid., p. 86.

27. Julia B. Hall, "History of C. M. Hall's Aluminium Invention," pp. 1-4.

28. Edwards, The Immortal Woodshed, pp. 83-5; Hèroult arrived at his invention by a route quite different from Hall's, noted to some extent in Martha M. Trescott, Rise of the American Electrochemicals Industry, chapter 3.

29. Charles M. Hall vs. P. L. V. Hèroult [here Hèroult's third initial is given as "V.", whereas many sources give it as "T."], in Interference in the United States Patent Office, October 24, 1887, typed document supplied the author by Mrs. Lydon of ALCOA; for Julia's testimony, see pp. 5-8 with the documentation of the 10th of February appearing on pp. 5-6.

30. Ibid., p. 8.

31. Ibid. For the testimonies of Heman Hall and the other witnesses, see pp. 8-12, with Heman's quoted statement here appearing on p. 9.

32. Ibid., p. 9; for the account of Hall's having shown the globules to Jewett on February 23, 1886, see Julia's testimony, p. 7, and also Charles's testimony, p. 4.

33. Ibid., p. 3.

34. Edwards, The Immortal Woodshed, p. 85. This episode deserves further and better documented study. Julia knew well that "'helping our lawyer' ... con-

stituted a fair day's work," as she noted on the
envelope of the February 10, 1889 letter from
Charles to her.

35. Edwards, The Immortal Woodshed, p. 226.

36. Carr, Alcoa, p. 18.

37. Edwards, The Immortal Woodshed, pp. 176-7.

38. "Aluminum Company of America Stock Ledger for Edie
    M. Hall, 1909-1919," "Aluminum Company of
    America Stock Ledger for Louie A. Hall, 1909-
    1925," "Aluminum Company of America Stock Ledger
    for Julia B. Hall, 1909-1925," plus a tabulation of
    "Dividends Paid on Common Stock," 1895-1943, all
    supplied the author by Mrs. Anna Lydon, archivist,
    ALCOA. For statements concerning the par value
    of ALCOA stock, see Carr, Alcoa, pp. 26-7, for
    the early days of the company; and Edwards, The
    Immortal Woodshed, pp. 112-3, for the early days,
    and p. 210 for post-1900 (early 1900s).

39. Edwards, The Immortal Woodshed, p. 177, for example.

40. Ibid., pp. 177-8.

41. Ibid., pp. 68-70 and 177-8.

42. ALCOA stock ledgers for Edie, Louie and Julia Hall
    (cf. note 38).

43. Charles M. Hall to Julia B. Hall, letter, November
    16, 1900, p. 1 of letter.

44. It is unclear what became of Heman Hall's shares when
    he died, and it would appear from the ALCOA stock
    ledgers that he did not will them to his daughters,
    even though Julia had taken care of him in his old age
    and decline. (See ALCOA stock ledgers and Oberlin
    Alumni Necrology for Julia Brainerd Hall and for
    Heman B. Hall.) For information on Edie and Louie
    Hall, see Oberlin Alumni Necrology.

45. Calculated on the basis of Julia's possession of 1,200
    shares of ALCOA stock by 1909, the most she
    acquired in her lifetime; the total dividend per year

rose to $6 in November, 1914, from $4 previously. Her average annual income from this stock after Charles's death until her own was approximately $8,000-9,000, when she possessed mostly 1,000 shares.

46. "The Perkin Medal, Remarks in Acknowledgment by Mr. Hall," Industrial and Engineering Chemistry, III (1911), 146.

47. Ibid., pp. 146-8.

48. This fact, of Charles's failure to credit Julia in this address, provoked many conjectures after the presentation on the Halls at SHOT in 1975, as to why he did not mention her. Also, the author is indebted to Professor Susan J. Kleinberg for raising this question and thereby helping to concretize some of the notions presented.

49. Both Carr and Edwards are motivated to comment on Charles's personality quirks at various points. For example, see Edwards, The Immortal Woodshed, p. 174, where it is stated, "To many of his co-workers, Charles was a peculiar man." See also pp. 202-3, which explains more about his seclusive and parsimonious nature. For his near-fetish concerning the Encyclopaedia Britannica, see pp. 164-5. It is not known how accurate Edwards' report of Hall as a boy is, but he claims that Hall was a kind of recluse as a reader, even as a young boy, as on pp. 2-3. Of course, it should be noted here that Hall made his famous discovery at the early age of 23, beginning serious experimentation at least by age 18. So he likely would have had to have been an intense student for some time previous to
    Carr discusses some of Hall's peculiarities on p. 20 of Alcoa, An American Enterprise. Also, one particularly interesting story about how Hall's acquaintances at Niagara Falls viewed him can be found in Edwards on pp. 208-9, and Edwards notes how both at New Kensington and at Niagara Falls, Hall tended to seclude himself in the lab when he got the chance, as on p. 182 and in Chapter 11.

50. Charles M. Hall to Julia B. Hall, letter, December 28, 1902, pp. 1-2 of letter.

51. "The Perkin Medal," pp. 146-8 for Hall's remarks and
    p. 149 for Hèroult's.

52. See, for example, photographs opposite title page of
    Edwards, The Immortal Woodshed, and on p. 8,
    p. 86, p. 136, p. 203, p. 225. Hall's extremely
    boyish appearance was the subject of comment
    throughout his adult life.

53. As the technical details in Charles's letters to Julia
    grew fewer after 1891, his technical correspondence
    with Hunt and Arthur Vining Davis grew larger,
    perhaps understandably. This can be seen in the
    ALCOA collection of Charles Hall's correspondence.

54. Alfred Hunt's mother, Mary Hanchet Hunt, encouraged
    Hunt to go to M.I.T. to study chemistry and metal-
    lurgy. She herself was a chemistry teacher at
    Patapsco Institute at one time. (Edwards, The Im-
    mortal Woodshed, p. 91; for a more detailed de-
    scription of Mrs. Hunt, see Junius D. Edwards, A
    Captain in Industry, 1957.)

55. On Charles Hall's attachment to Charles Chandler, see
    Carr, Alcoa, p. 20. Also, Dr. Chandler said in
    1911 in introducing Hall as Perkin Medal recipient,
    "Personally, Dr. [sic] Hall is a man of the greatest
    modesty, and most lovable character." ("The Perkin
    Medal," p. 146.)

56. Compare Edwards, The Immortal Woodshed, p. 119
    with the letter from Charles to Julia, May 5, 1889,
    p. 1 of letter.

57. See, e.g., letters from Charles to Julia, December 28,
    1902, pp. 1-2 of letter for reference to the "girls";
    also pp. 85-6 of Edwards, The Immortal Woodshed,
    where Edwards correctly cites the April 8, 1889,
    letter, p. 4, which shows Julia's having complimented
    Charles's work after he received his patents. See
    letter, Charles to Polly (a younger sister), June 20,
    1889, in which Charles indirectly compliments both
    this sister and Julia on their looks, p. 1 of letter.
    Finally, see letter, May 25, 1891, in which Charles
    tells Julia that an old friend of his, Dwightie Bald-
    win, is getting married and he (Charles) asks Julia
    what kind of gift or congratulations to send. Since,

he says, "you have had such a great amount of experience in such matters and are well known to be
a walking encyclopedia on social usages, I write to
you to advise me what to do" (p. 1 of letter).   This
last one is credit to Julia in some areas, but not in
the matter of the aluminum invention and related
business and technical events.

58.   See Elizabeth Gould Davis, The First Sex (New York,
      1971), p. 305 and Edwards, The Immortal Wood-
      shed, p. 19.

59.   Margaret Rossiter, "Women Scientists in America Be-
      fore 1920," included in this anthology.   Eleanor
      Flexner in Century of Struggle, the Woman's Rights
      Movement in the United States, 1974, gives an il-
      luminating account of the progress of women in
      science and medicine in the eighth chapter, which
      also includes information on the progress of women
      in higher education generally.   Also, see Edythe
      Lutzker's Women Gain a Place in Medicine (New
      York, 1969).   For sketches of various types of
      female scientists in U.S. history, see Notable
      American Women, a Biographical Dictionary.   In
      addition, this author's own research on female
      electrochemists and inventors at the turn of the
      century has uncovered instances of ridicule of the
      women, as on p. 395 of the Electrochemical Industry,
      I (1903), in a description of Mary Emme's U.S.
      patent for a storage battery.   The fact that this in-
      vention is somewhat ludicrous is not as significant,
      it would seem to this author, as the amount of space
      the editors devoted to embellishment upon this fact.
      Surely a huge percentage of all patents, regardless
      of country granting the patent, could be considered
      fantastical and laughable.   The interesting thing is
      that the journal editors so singled this one out and
      devoted as much space as they did to ridiculing it.

60.   Alva Matthews, "Some Pioneers," speech presented be-
      fore the Society of Women Engineers, n.d.   A copy
      of this document was supplied the author by Ms.
      LeEarl Bryant of the S.W.E. in 1976.

61.   Charles M. Hall to Julia B. Hall, letter, February 10,
      1889, p. 1 of letter.

62. Charles M. Hall to Julia B. Hall, letter, May 3, 1891, p. 1 of letter.

63. See Edwards, The Immortal Woodshed, p. 27, which correctly quotes the letters in question.

64. Raymond Szymanowitz, Edward Goodrich Acheson, Inventor, Scientist, Industrialist (New York, 1971), pp. 106-7, 115, 124, 151, 258-291 for various descriptions and quotations concerning Margaret's own technical background, her assistance of Edward in his technical and business pursuits, and their relationship. Also, it should be noted that the Acheson papers are held at the Library of Congress, and the author obtained photocopies of much of these. However, the author's file of copies holds no personal correspondence between Edward and Margaret Acheson. The author is indebted to the assistance of Raymond Szymanowitz, Robert Vogel, Diane Newell and Antoinette Lee in obtaining portions of the Acheson file.

65. For commentary on Zula Sperry, wife of Elmer, see Thomas P. Hughes, Elmer Sperry, Inventor and Engineer (Baltimore, 1971), especially pp. 98-9 and 101.

66. This approach to Edison has most recently been stressed by Thomas P. Hughes in a presentation before the Society for the History of Technology, October, 1975, summarized in Technology and Culture, XVII (July, 1976), 502 and also in Science, 190 (November 21, 1975), 763. This paper appears in full as "The Electrification of America: The System Builders," Technology and Culture, XX (1979), 124-161. For Edison, see especially pp. 125-139.

67. Edwards, The Immortal Woodshed, p. 85.

PART II

EFFECTS OF TECHNOLOGICAL CHANGE ON WOMEN IN THE DOMESTIC SPHERES

INTRODUCTION TO PART II(A)

Women as Housewives and Homemakers

As Part I closes on the Halls and the Pittsburgh Reduction Company, Part II opens on another essay dealing not only with an aspect of Pittsburgh's history but also with essentially the same time period as the Hall piece. Also, the essay on Julia Hall focused on the contributions of female relatives of inventors, such as the sisters and wives. Part II(A) now views the wives and technological change in yet another light: the changes in domestic and municipal technologies which affected housework. Susan J. Kleinberg has chosen to deal with the late nineteenth century, Ruth S. Cowan has focused on the twentieth. Kleinberg is mainly concerned with working class women, Cowan deals more with the middle-class housewife.

Kleinberg makes the point that changes in domestic and municipal technology, such as, in the former, washing machines, and, in the latter, the spread of sewers, public water supplies and paved roads, did not reach the working classes nearly as quickly as the middle and upper classes. Therefore, the lives and labors of the working-class housewives were still quite onerous compared to their wealthier counterparts by 1900. Her piece suggests that as we view technological change in the household, we must differentiate among economic classes, as the rates of diffusion of these technological changes were often much slower for the less well-to-do.

Ruth Cowan's essay in some ways takes up where Kleinberg's ends. While the essay by Cowan is valuable for its sheer information on changes in the technology of the household and housework, it also postulates an important thesis: namely, these technological changes actually reduced the status of the middle-class housewife, who increasingly performed more of the housework herself, aided by fewer and fewer domestic servants. Where once she had been a manager of various domestic servants, machines displaced these servants, resulting in the proletarianization of the housewife. Further, Cowan has found that for those who were full-time housewives, the amount of housework they did actually increased during the twentieth century with the advent of more sophisticated household technology. The traditional view that such technology must have actually provided the housewife with more free time is challenged by Cowan's work.

1. TECHNOLOGY AND WOMEN'S WORK: THE LIVES OF WORKING CLASS WOMEN IN PITTSBURGH, 1870-1900*

by Susan J. Kleinberg

> It is through the households themselves that the industrial situation impresses itself indelibly upon the life of the people[1]

Most examinations of working class women focus on wage earning women in textile towns and in mixed industrial and commercial centers.[2] However, historians need to expand their examination of working class women to include those living in cities which had few employment opportunities for women such as Pittsburgh, and to study women's unpaid work in the home as well as paid labor outside it.[3] The work women did in their own homes for their own families is as important and worthy of historical analysis and as much a contribution to the economy as was work done outside the home for cash wages. This work done in the home enabled families to manage in an industrial, urban setting in which the benefits of industrialization and urbanization were distributed unequally. If we are to understand the fabric of working class life in Pittsburgh or any other mill or mining town during the decades which followed the Civil War, indeed, if we are to understand the lives of the working class at all, we must look at the lives of the women as well as those of the men, in the households as well as in the workplaces.

A study of domestic and municipal technology, who benefitted from it, and who could not afford it illustrates the deprivation of the working class vis-à-vis other groups in

---

*I am indebted to Professors Herbert Gutman, Anne Firor Scott, and David Montgomery for their comments on successive drafts of this paper.

This article is reprinted by permission from Labor History, Winter 1976, pp. 58-72.

Mill Workers' Houses and Privies, Sylvan Avenue, Pittsburgh, 1907. Photo courtesy of Carnegie Library, Pittsburgh.

the city and shows how urbanization and technological innovation affected working class lives. Such a study provides
a microcosm of governmental attitudes toward the working
class and shows the ways in which urban government could
improve or ignore the quality of life of different socio-eco-
nomic groups within the city.

Given the underlying philosophy of government and
private enterprise in Pittsburgh at the end of the nineteenth
century, namely that those people who benefitted from services should pay for them, it followed that the distribution of
municipal services and technological innovations proceeded
along class lines and was influenced by political considerations. Those who could pay got, and those who could not
did without. This was as true in the public sector as in the
private, since property owners were expected to pay for the
services provided to their property. Although operating and
maintenance expenses were borne by the city treasury (i.e.,
the tax-payers as a whole), the actual cost of original street
and sewer improvements was assessed against the property
owners "abutting and directly affected, to the extent benefitted.... "[4]

Since people in some neighborhoods were unable to
stand the cost of the improvements, they did without. Conditions were worst in working class neighborhoods where
the majority of the residents were tenants rather than property owners. The general shortage of housing in Pittsburgh
meant that landlords did not need to provide services to
entice tenants. As a result, working class neighborhoods--
aggregates of tenant-occupied buildings and tenement areas--
had fewer amenities than did more established areas. They
had fewer sewers, fewer paved streets, and less water. The
lack of services in working class neighborhoods was all the
more critical since these areas were precisely the ones
which suffered most from the pollution of the mills and had
the most traffic.

Pittsburgh's geography effectively made working class
neighborhoods dirtier, less pleasant places. The workers
and their families lived in cramped housing on the hillsides
above the mills strung out on the narrow flat lands along
Pittsburgh's two rivers. The steep hills, the lack of paved
roads, and the railroads which served the mills but not the
people living on the hills all effectively isolated working
class communities from the rest of the city. By contrast,
the middle and upper classes moved to gently sloping suburban

areas, served by street cars and commuter railroads, far from the din and dirt produced by the mills.[5] The continued maldistribution of municipal services reinforced the segregation and differences in life styles between the classes.

The variations in street paving in two sections of the city, the wealthy suburban "East End" and the working class "Point" district highlight the political nature of municipal services and show how political decisions reinforced existing social differences. In 1873, the Pittsburgh Board of Health characterized the Point as "without exception, the filthiest and most disagreeable locality within the limits of the city." The Board of Health asserted that paving the streets of the Point would enhance the general health of the city. The landlords, however, maintained that the tenants had to pay all the expenses of any improvements as a requirement of their leases. Since the tenants could not afford to do so, the Board of Health suggested that the city bear the cost. The City Council refused to act on the suggestion, and the streets remained unpaved into the next decade.[6]

All improvements had to be paid for thirty days after completion of the work. The one exception to this rule occurred in the newly settled suburban East End, the commuter suburb to which the industrialists moved in the 1860s and 1870s. Since improvement costs were assessed on a per-frontage foot basis the large estates of Mellon, Frick, and other industrial magnates would have paid heavily for the roads which made their commuting possible. According to the City Controller, paying the costs within the specified time would have been "a serious burden on the owners of the property assessed for the improvements." In order to relieve this burden, "a prominent and public spirited citizen" of the East End proposed, and the City Council agreed, that the city should sell bonds to cover the costs of paving and curbing the streets. The costs would be recovered by assessing the property owners in ten equal annual installments, rather than the single payment required of inner-city lot owners. The act itself was deemed unconstitutional, and the city assumed the entire cost of paving the suburban roads. It did this during the same years that it refused to pave the streets down at the Point.[7]

The increased mobility which paved streets (and the surface transportation which used them) provided were also distributed unequally. Women who lived in the suburbs could take the street car into town. Those living in working class

neighborhoods without paved roads and therefore without
street cars could not.[8]   Their access to stores was limited.
Their mobility did not expand as the city grew.   The taxes
which their rent money paid paved the streets to the suburban sections.   If their own streets were paved at all, the
costs were assessed against their landlords, who passed
them on by raising the rent.

As the city became more densely populated, working
class families suffered as a result of unsanitary conditions
which, in turn, resulted from governmental decisions as
well as from over-crowding.   Unpaved roads, for example,
could not be kept in good order.   It was more difficult to
sweep or clean them.   Horse droppings and human refuse
remained there longer, putrefying in warm weather.   Moreover, the installation of sewer and water pipes did not keep
pace with urban expansion in working class neighborhoods.
These were slighted while the pressure on the cesspools and
inadequate water pipes increased.   People living in working
class neighborhoods were exposed to the disease these conditions fostered, epidemics struck them most heavily, and
death rates rose dramatically in new industrial areas.[9]

Indeed, probably the most striking and significant
examples of unequal distribution of municipal services were
the allocation of water and sewers.   Working class neighborhoods had less water and they had it in a less convenient
form.   While middle class homes were serviced by large
water mains and indoor water pipes, working class homes
were served by smaller water pipes with pumps in the courtyard or down the street.   The discrimination in pipe size
resulted from a City Council Water Commission decision
made in 1872, to lay pipes in accordance with the amount of
revenue each street could be expected to produce.   The rationale for this unequal distribution was that it would not
deprive "the paying portion of our city of a sufficient water
supply. "   This cost conscious method of allocation meant that
some streets did without water, others had an insufficient
supply, and a few could use water with impunity.   Here the
city's geography played a part.   The city's equipment did not
generate sufficient pressure to push water up the steep hills
in the summer when the reservoir was low.   This meant
that the poor

> living in remote areas which are reached by small
> service pipes only, are subject in [hot] weather
> when cold water is their greatest boon, to actual

> suffering for the quantity for which they pay and
> which is essential to health, cleanliness, and
> comfort. [10]

The problem lay both with the smallness of the pipe
and with the great demand from the industrial users located
on the flat lands below the working class neighborhoods.   In
warm weather, when the water supply was limited and the
pressure lower, the mills and railroads diverted the water,
leaving the hillsides dry above them.   For example, the
Pennsylvania Railroad supplied its locomotives with water
from a pipe at the bottom of a hill.   In so doing it cut off
the supply to the railroaders and mill workers living above
the yards.   During the summer, many neighborhoods in the
largest industrial section of the city (the South Side) had no
water from seven in the morning until six at night while the
mills operated. [11]

The burden of unequal water distribution fell most
heavily on working class women.   They had to take great
precautionary measures to have enough water to last throughout the day.   If South Side women wanted water for their
housework or to drink, they hauled it all early in the
morning before the mills started up.   One woman complained that in order to have water during the long summer
days, she got up at five o'clock in the morning to fill her
tubs.   The uncertainty of the water supply in working class
neighborhoods made women's household chores harder and
interrupted their routine.   Even the daily newspapers recognized that there was "nothing more calculated to disturb the
serenity of the household than to have the water supply
turned off on wash day."[12]

Not only did the working class have less water, but
they had to bring it indoors by hand.   Although by 1890,
the Chief of the Department of Public Safety claimed that
it was the "custom to have even moderate priced houses
fitted with hot and cold water, water closets, and bath
room," such amenities were beyond the reach of the working
class.   They could not afford the rents in houses which had
indoor water, they could not afford to install the plumbing,
nor did they always live near the necessary sewer and water
lines.   Moreover, the water companies (one public and one
private) charged more for fixed indoor plumbing than for a
pump in the yard.   An indoor toilet cost $2.50 more per
assessment period, a fixed sink cost $2.00 more, and a
fixed tub cost $4.00 more.   The water companies also

discriminated against smaller householders. They charged about twice as much per room for water in smaller dwellings as they did for the larger ones. Since there were no water meters, those with indoor water could afford to use it lavishly. A Department of Public Safety report declared in 1889 that "a walk through the wealthy or business portion of the city will show the small regard they have in the use of water or the welfare of others." Water was a necessity; having it indoors was a luxury the working class could not afford. [13]

As a result, working class women carried indoors "every drop of water they would use." If they lived on the second or third floor they carried the water upstairs and back down, disposing of it in the sink in the yard, on the ground, or sometimes out the window, but always hauling water. Day after day and many times a day, a woman living in a mill tenement "carried water up and down that her home and her children might be kept decent and clean ... Her shoulders and arms had to strain laboriously...." This woman and all working class women had to carry water inside for cooking, for cleaning, and for washing dishes and clothes. No matter what the chore, she needed water to do it. The political decision of the city to lay small pipes in poorer neighborhoods, the great demand from the mills and railroads, and the precarious economic situation of the working class which resulted in the location of water outside rather than inside the home all made her housework more complicated. Her washing and cleaning chores, made difficult by Pittsburgh's heavy particle pollution and by the grime and sweat on her family's clothes, were made more arduous by the city's decision to provide decent services only to those who could pay for them. [14]

In addition to the high cost, inconvenience, and scarcity of water, the working class residents of Pittsburgh's South Side (its largest industrial community and almost entirely working class in composition) suffered from the poor quality of the water itself. The South Side was served by a private water company, the Monongahela Water Company, originally established to provide water for the industrial establishments along the Monongahela River. According to the American Glass Review, the newspaper of the glass workers, the water was "drawn from the Monongahela River, just below where the large sewers discharge their filth into the river." The Monongahela was the catch basin for the refuse and sewage of the more than 60,000 people living along its banks. It also carried the refuse from mills, glass houses,

Tenement Court, Pittsburgh, c. 1900.
Photo courtesy of Carnegie Library, Pittsburgh.

and slaughterhouses which stood on both sides of the slow-moving river.[15]

The city periodically protested the poor quality of water provided by the Monongahela Water Company. The only action it took during this era, however, was to run water pipes to the South Side from the public water company to insure sufficient water in case of a fire. "... The babe, the mother, the robust young man, and all who quench their thirst from the supposed, and should be, pure beverage ..." declared the American Glass Review, continued to drink sewage-contaminated water. And the mill workers' on-the-job drinking water came directly from the river with no pretense of filtration or purification, the pipes being located directly down-stream from the point at which the dump boats emptied sewage. The men who worked in the mills, some of whom drank as many as thirty or thirty-five glasses of water during a shift to prevent dehydration from the intense heat, consumed water freshly contaminated by sewage, refuse from other mills, and wastes from the slaughterhouses.[16]

In the late nineteenth century, sanitary conditions in working class neighborhoods did not improve as they did in other areas of the city. Working class wards, for example, suffered from a relative lack of sewers in proportion to the density of their populations. In 1879, the most extensive sewer connections were in the fashionable parts of the city. The Central Business District and the heavily upper class Fourth Ward had excellent sewer connections. The Point had no sewers whatever, nor did large sections of the working class wards along the rivers and on the South Side.[17]

Ten years later, in 1889, although the Chief Sanitary Inspector described the city as generally well-sewered, parts of the South Side still had no access to sanitary waste disposal. Several hundred cesspools drained into abandoned coal mines, primarily on the hill tops above the mills and railroad yards. Since these areas were without paved streets, the expense of transporting sewage to a proper dumping ground and the large outlay necessary to provide proper sewage "led many to resort to the very dangerous practice of drilling their privy wells and cesspools through the comparatively thin crust of earth overlying these coal mines." Since many of the mines had caved in, the sewage was trapped there, seeping into wells upon which some South Siders still relied for their water, and creating a particularly bad aroma.[18]

All such conditions made life in working class neighborhoods decidedly less pleasant than in middle and upper class ones. They also resulted in more labor for working class women. The hard physical work their husbands and children did meant sweaty, dirty clothes which had to be washed by hand with a sporadic and inconvenient water supply. Although the women themselves tried to manage, it was an uphill struggle:

> Where the yards are paved, women may be seen at any hour of the day hard at work with their brooms on an effort to keep the place clean. But in many cases there is no such paving and the water and other fecal matter soaks into the rotten planks that have been spread about the courts by the women in an attempt to keep the mud from being tracked into the house. [19]

Beginning in the 1870s some American households were revolutionized by mechanical devices capable of doing housework (e.g., washing machines, central heating, toilets, and iceboxes), the replacement of dirty fuel (coal and wood) by clean fuel (gas and electricity), the city's provision of paved roads, sewers, and municipal water systems, and by new forms of communication and transportation. Well into the twentieth century, however, the benefits of the new municipal and domestic technologies were limited to the middle classes and the very upper reaches of the working class. The depressions which blighted these decades, in 1873-79, 1883-86, and 1893-97, kept most of the working class too poor to afford them. As a result, the domestic burdens of working class women were not eased by the new domestic technology which made middle class homes pleasanter and more comfortable. Instead, working class women continued to do housework in much the same manner as their mothers did, although the city became dirtier, less sanitary, and more crowded as a result of increased industrialization and urbanization. [20]

In a working class society attached to industries which relied primarily on male labor, men worked outside the home and women worked within it. Each member of the family pulled her or his own weight by fulfilling the allotted roles; earning money, or managing the family economy and domestic production. [21] In such an environment, a woman's economic contributions differed fundamentally from a man's, but were still important to her family and city. The hard physical labor a woman did in her own home enabled the family to manage on the husband's (and sometimes children's) income. In times of economic duress, frequent during this era, it was the wife's responsibility to make ends meet on the reduced pay packet. She decided how the family would manage, cooking cheaper food, economizing in the household wherever possible. One form this economy took was the continued use of the housewife's own labor to perform chores which were taken over by machines or done by servants in middle class households. This was both an accommodation to the industrial system, with its frequent periods of unemployment and depressions, and an indirect support of the urban industrial structure which deprived the working class while providing for the middle and upper classes. The term support must be used advisedly, of course, for working class women no less than their husbands wanted fatter pay packets.[22]

Since married working class women in Pittsburgh did not work for cash wages, they depended upon their husbands (and to some extent their children) for support.[23] They were managers rather than earners and without an independent say in the allocation of the family income. They received whatever portion of their husband's pay packet he chose to give. Implicitly, women did not have the right to know how much their husbands earned or how much he kept for himself. The author of one of the first standard of living studies among working class families noted that

> the small expenditure for tobacco and liquor in these budgets is to be accounted for, at least in part, by the fact that men did not tell at home what they had purchased. The women usually hesitated to ask the man about his spending money, and as in the days of slack work they did not know just what he earned.[24]

It was a woman's job to manage on whatever money her husband gave her, regardless of the amount taken before he turned over his pay packet to her. Working class women's domestic labor and ability to work under crowded home conditions made bearable their husbands' low wages and bore the brunt of whatever drinking, smoking or spending habits he might have. This is not to say that all or even most working class men had such habits, but that where they existed it was the wife who had to compensate--feeding, clothing, and housing the family, regardless. The wife may have been the chief spending agent, but she spent within the constraints set by the low wages and her husband's taking his spending money first.

If the findings of sociologists are adequate evidence, another result of married women's exclusion from the labor force was a lack of power within the family. Studies done in the United States in the 1930s, 1940s, and 1950s indicate that when women worked they had more influence over "really important decisions" than women who did not work outside the home. Work inside the home was not regarded as an economic contribution to the family, and therefore, house-wives "had little basis for demanding a large voice in major economic decisions."[25] In that way, although their work at home sustained the family and made possible its getting by in the face of reduced income and inadequate municipal serv-ices, the work itself was unvalued or undervalued by the working class family.

There were two factors which kept working class women tied to the older, more laborious methods of doing housework at a time when other women substituted machine power for human power. The first and most significant was the scarcity and irregularity of working class employment and the low wages received. The new household technology required a larger capital investment than most working class families felt they could afford. Even "the decision to buy a new piece of furniture is often a matter for grave considera-tion." In a period of economic unrest, when working class families kept expenditures to a minimum, they could not af-ford such expensive items. Moreover, while they eased the women's work load and enhanced their self-esteem, they were not necessary for family survival, as were food, housing clothing, and accident insurance. Although "housework may be materially lightened by the use of gas instead of coal," it was too expensive for working class families.[26]

The second factor which kept working class women from obtaining consumer durables was that husbands did not appreciate the amount of work done by their wives. In one woman's words, "the only time 'the mister' notices anything about the house is when I wash the curtains." Men con-sidered it natural for women to labor at home. They did not, therefore, see the need to replace their wives' physical labor with machine power. A contemporary observer noted that "men seldom appreciate the worth of a devoted wife and look upon her many kind acts as a matter of duty." Because men were away from the home when their wives did most of

the housework, and came home tired from their own work,
and because society took women's work in the home for
granted, it followed that when money was scarce, working
class women would continue to do their housework in the old
way. Replacing their labor with machines did not have a
high priority in the family budget. [27]

The economically deprived working class women still
brought water in from the pump in buckets to heat on a coal
or wood stove. The stoves themselves needed tending; lamps
had to be kept trim and clean. Ammonia, cleanser, soap
and water, rags, dusters and the all-important elbow grease
were the only tools working class women had to combat the
grit, dust, and dirt from the mills. Middle class women increasingly the medium through which their husbands displayed
affluence, had indoor hot and cold running water, gas stoves,
lighting and heating, washing machines, telephones, and, of
course, servants to do the heaviest work. [28]

What the new household appliances did, obviously,
was to substitute mechanical energy for human energy. That
meant housework became less physically arduous and that
housewives gained some measure of free time. For example,
the forced hot air furnace, which provided even, dirt-free
heat to the entire house, was a great advance over the fire-
place, which needed frequent tending and did not heat a whole
room, and the closed stove, which did heat a room but still
needed to be fired up, fed, and cleaned. A furnace cost anywhere from $75 to $300, while closed stoves cost $10 or
$11. Most working class homes were heated by closed
stoves, although some in the tenement districts still relied
upon fireplaces. The cost of owning and operating even one
such stove was high enough so that most working class homes
had heat in only the kitchen. Photographs of working class
homes taken during this period show a single chimney in the
rear of the house. Middle class homes, on the other hand,
had chimneys front and back and on both sides. These homes
were heated with clean gas rather than the coal and wood
common to working class homes. [29]

As with heat, gas was substituted for the dirtier forms
of energy in the cooking and lighting of middle class homes.
The gas stove was cleaner, cooked more evenly, and, most
marvelous of all, did not heat the entire kitchen to insuffer-
able temperature during the summer. Cooking became easier
and the kitchen a more pleasant room in which to work. The
pleasantness of the kitchen is important, since it was the room

in which a working class woman spent most of her time.
She cooked, washed clothes and ironed there.

Meal preparation itself continued to take a large portion of the working class woman's day. Marketing had to be
done almost daily. The cost of an icebox and ice were high.
(Ice alone cost forty-two cents per week for the minimum
amount.) The low wages of the working class, especially
laborers, prevented the purchase of items in the larger eco-
nomical quantities. This was a severe problem in Pittsburgh, which had very high food prices. For example, a
barrel of flour which would last the average family for six
weeks, cost six dollars, a large amount to save for families
living on seven to ten dollars a week. The irregular work
schedules of her husband, children, and possibly boarders,
meant that the working class woman was cooking constantly.
Some men complained bitterly if their meals were not ready
when they came home and physically abused their wives for
this outrage. Others demanded that their wives serve them
no matter what the circumstance: whether they came home
late or if their wives were sick. The demands of meal
preparation coupled with other household chores put cease-
less pressure on working class women. That these women
had more work than they could readily manage was made
clear by one woman, who felt that a good husband was one
who was careful of his clothes, didn't drink, and would "eat
a cold dinner on washday without grumbling."[30]

Washing clothes was certainly the most arduous and
least pleasant of household chores. Virtually all working
class women did their own laundry. The steam laundries
which appeared in Pittsburgh in the mid-1880s catered to the
affluent and to single men. Working class families sent
their laundry out only if the wife was sick and no daughter
or other relative could do the wash for her. Washing machines cost $15, a week's salary for skilled workers and
two weeks salary for laborers. Even after the turn of the
century, washing machines were limited to working class
families with "larger incomes," the aristocracy of the iron,
glass, and steel trades.[31]

Working class women did their laundry in the old-
fashioned way. They brought water in from the pump,
heated it on a coal or wood stove, emptied it into a wash-
tub, and scrubbed the clothes while bent over the "back-
breaking washboard." Soapy water was carried outside and
emptied, clean water brought in from the pump once more,

heated on the stove, and used for rinsing the clothes. Especially sweaty and dirty clothes, those worn by the men in the mills or children's play clothes and soiled infantwear, might well be soaped and rinsed again, then wrung out and hung to dry. In winter, washing was done in the kitchen. During the warm weather it might be done on the back porch if the family had one, in the courtyard or alley, or in the overheated kitchen. Ironing was also done in the kitchen, near the stove on which the iron was heated. Washing and ironing were hard, hot work, done under cramped conditions.[32]

While the washing machine was only one of many labor-saving devices available during this era, the limits of its diffusion were indicative of the dispersion of technological advances. Both public and private technology were distributed according to the ability to pay. The notion that what happened in the working class neighborhoods affected the rest of the city had not yet taken hold in Pittsburgh.[33] The purchase of domestic technology was the purchase of leisure time for the women of the household. This technology-for-hire system had the effect of concentrating working class women's roles in comparison with the expansion of middle class women's roles during the latter decades of the nineteenth century. It also brought into sharp relief the distinctions between middle class and working class households. Where the working class woman had worked in a middle class home and had been exposed to the marvels of flush toilets and washing machines, the contrast must have been great. As Margaret Byington pointed out in her pioneering study of steelworkers' households:

> We are all imitators, and the inability to have what others have, even when the absence of the thing is not in itself a privation, reacts on the individual life by lessening the sense of self-respect and social standing. [34]

It was not only that conditions did not improve in the working class households, but that they improved so dramatically in middle class ones.[35] At the time when increased traffic, crowding, and industrial pollution made cleaning difficult for working class women, middle class families moved away from Pittsburgh into paved, sewered, suburban neighborhoods and obtained the appliances which made housework easier.

The economic structure of the city prevented working class women from making the cash contributions to their families

which might have increased their power and broadened their
role within the family. In effect, their labors in the home
and their exclusion from the labor force permitted and reinforced the traditional role segregation of women and men.
At the same time, the unequal spread of municipal and domestic technology meant that these women continued to spend
long hours in the home. The long term effect of the government and private enterprise pay-as-you-go philosophy was to
heighten class and sex role differentiations. In Pittsburgh,
where women could not work outside the home, it also made
women into indirect participants in the industrial system, for
it was their labor and ingenuity which sustained the working
class family throughout this period. Indeed, their unpaid
labor helped sustain the entire urban industrial system.

Notes

1. Margaret Byington, Homestead, the Households of a
   Milltown (New York, 1910), p. 179. This pioneering
   study of women's work in the home in an industrial
   town strongly influenced my thinking about women's
   work in a heavy industry environment.

2. Hannah Josephson, Golden Threads, New England's Mills
   Girls and Magnates (1949); Ralph Scharnau, "Elizabeth
   Morgan, Crusader for Labor Reform," Labor History
   XIV (Summer, 1973); James Kenneally, "Women and
   Trade Unions, 1870-1920," Labor History XIV
   (Winter, 1973); Alice Kessler-Harris, "Between the
   Real and the Ideal, Studies of Working Women in 19th
   and 20th Century America," paper read at the Or-
   ganization of American Historians, 1974.

3. I am using the term working class synonymously with
   manual worker, no matter how skilled. In periods
   of industrial chaos there was a high rate of unemployment and even the most skilled workers suffered.
   In Pittsburgh, one-third of all working class men
   were unemployed for at least one month in the year
   preceding June, 1880. On the differences in the
   standard of living between unskilled and skilled
   workers see U.S. Congress, Senate Committee on
   Education and Labor, Report upon the Relations Between Labor and Capital (Washington, 1885), v. 1,
   pp. 17-31, the testimony of Robert D. Layton, Grand
   Secretary of the Knights of Labor of North America.

4. Pittsburgh Public Works Dept., The City of Pittsburgh
    and Its Public Works (Pittsburgh, 1916), p. 19;
    Pittsburgh City Controller, Report (Pittsburgh, 1895),
    pp. 12-13; Pittsburgh Board of Health, Annual Re-
    port, 1873 (Pittsburgh, 1874), pp. 19-20.

5. Joel Tarr, "Transportation Innovation and Changing
    Spatial Patterns: Pittsburgh, 1850-1910." Urban
    Mass Transportation Administration, April 1972,
    p. 14.

6. Board of Health, 1873, pp. 19-20.

7. Pittsburgh City Controller, p. 16.

8. The street cars ran on the major paved roads which
    connected the new suburban areas to the Central
    Business District. See Tarr, passim.

9. Pittsburgh Board of Health, Annual Reports, 1873-1887.

10. Pittsburgh Public Works Dept., p. 19; Pittsburgh Water
    Commission, Report, 1872 (Phila., 1873), p. 5;
    Superintendent of Water Works, Annual Report, 1877-
    1878 (Pitts., 1878), p. 5.

11. Superintendent of Water Works, pp. 3-4; Pittsburgh
    Commercial Gazette, June 19, 1888.

12. Pittsburgh Commercial Gazette, June 19, 1888, July 18,
    1877.

13. Department of Public Safety, Report, 1889 (Pittsburgh,
    n.d.), p. 78; Superintendent of Water Works, Annual
    Report, 1878-1879 (Pittsburgh, 1879), p. 5.

14. William Henry Matthews, A Pamphlet Illustrative of
    Housing Conditions in Neighborhoods Popularly Known
    as the Tenement House Districts of Pittsburgh
    (Pittsburgh, 1907), p. 5; Paul Kellogg ed., The
    Pittsburgh District (New York, 1914), p. 132.

15. American Glass Review, December 12, 1887.

16. Annual Report of the Mayor of the City of Pittsburgh to
    the Select Council in Minutes of the City Council,
    1884, pp. 127-28; Minutes of the City Council, 1885,

pp. 85-87, 212, 235; American Glass Review, December 12, 1887; Every Saturday, March 18, 1871, p. 263.

17. Board of Health, Annual Report, 1878-1879, pp. 20-23. There is a map of the sewer system of Pittsburgh appended to this report.

18. Crosby Gray, The Past, Present and Future Sanitation of Pittsburgh (Pittsburgh, 1889), p. 69; Department of Public Safety, 1889, p. 44, 1890, p. 66.

19. Paul Kellogg, ed., Pittsburgh District, p. 96.

20. The best general histories of the diffusion of household technology are Elizabeth Bacon, "The Growth of Household Conveniences in the U.S., 1865-1900" (unpublished Ph.D. diss., Radcliffe College, 1942), and Siegfried Giedion, Mechanization Takes Command, a Contribution to Anonymous History (London, 1948).

21. The conclusions about women's labor force participation are based upon a ten per cent sample of the manuscript census, Pittsburgh, 1880. Less than one per cent of all married women worked outside the home. Those who did so came from the working class and had such jobs as laundress or seamstress. One way in which women did contribute to the family income was to take in boarders. For a fuller explanation of the work involved in boarding see my diss., Technology's Step-daughters: The Impact of Industrialization upon Working Class Women, Pittsburgh, 1865-1900 (University of Pittsburgh, 1973).

22. Byington, pp. 38, 60.

23. Children, in fact, were the ancillary breadwinners in Pittsburgh. Byington, p. 41, and Kleinberg, ch. 5.

24. Byington, p. 82.

25. William Chafe, The American Woman, Her Changing Social, Economic and Political Role, 1920-1970 (New York, 1972), pp. 220-224.

26. Byington, pp. 85-86.

27. Ibid., p. 108.

28. This argument is drawn from Thorstein Veblen, The
Theory of the Leisure Class, an Economic Study
of Institutions. Although outside the focus of this
paper, it should be noted that middle-class women
were becoming cultural and political forces in
American Society. The General Federation of
Women's Clubs, Women's Christian Temperance
Union, suffrage and reform movements of the late
nineteenth and early twentieth century were middle
class movements with few, if any, working class
participants. See also Bacon, p. 48.

29. Bacon, pp. 25, 28-29, 219-220, 98-103. For pictures
of these homes see Pittsburgh District, The Civic
Frontage. Working class homes built during these
years sometimes had a hole cut in the ceiling to
allow heat to rise to the upper floor. This arrangement was typical of the homes along the railroad
tracks in Skunk Hollow near Boundary Way. Some
are still used as homes with space heaters and
plumbing added. The mansions of the era had
central heating and/or individual room gas fireplaces known as Taylor Burners. The fancier gas
stoves had "water backs," hot water heaters which
provided hot running water. Such stoves were expensive and required costly indoor plumbing. Having
a supply of hot water always on hand, however, made
housework much easier. While gas lighting was
found in most middle class homes, kerosene was used
in the "vast majority of dwellings among the common
people." (Bacon, p. 104.)

30. Pittsburgh Post, April 20, 1881; Price list for the
Chautauqua Ice Company. U.S. Department of Labor,
Bulletin, July 1907, pp. 175-328; Byington, pp. 66,
74-79; Pittsburgh Post, July 2, 1868, January 24,
1870; Western Pennsylvania Humane Society, Case-
work Records (Manuscript in possession of W.P.H.S.,
Pittsburgh), January 9, 1889; People's Monthly, June
1, 1872, p. 3.

31. The ownership of the first forty washing machines in
Pittsburgh was traced. All but four were clearly
middle class. These four were the wives of a
carpenter and an alderman and two women who kept

boardinghouses. Pittsburgh Gazette, January 14, 1870. The Dexter Washing Machine advertisement claimed that the machine was "cheap, simple, durable, effective" and did the "ordinary washing" of a family in only one or two hours. It saved soap and required no boiling of clothes.

32. Byington, p. 87; I am grateful to Mrs. Weir of Wright Street, South Side, Pittsburgh for her description of washing clothes when she was a child. She made the process graphically understandable to a person brought up in the age of the electric washing machine.

33. This notion paved the way for municipal garbage collection and the extension of the water sewage and road systems. It gained currency among urban reformers during epidemics but did not become widespread until the Progressive era. See Charles Rosenberg, The Cholera Years (Chicago, 1962).

34. Byington, p. 85.

35. Ibid., p. 79.

2. THE "INDUSTRIAL REVOLUTION" IN THE HOME:
Household Technology and Social Change
in the 20th Century*

by Ruth Schwartz Cowan

When we think about the interaction between technology
and society, we tend to think in fairly grandiose terms:
massive computers invading the workplace, railroad tracks
cutting through vast wildernesses, armies of women and
children toiling in the mills. These grand visions have
blinded us to an important and rather peculiar technological
revolution which has been going on right under our noses:
the technological revolution in the home. This revolution
has transformed the conduct of our daily lives, but in somewhat unexpected ways. The industrialization of the home was
a process very different from the industrialization of other
means of production, and the impact of that process was
neither what we have been led to believe it was nor what
students of the other industrial revolutions would have been
led to predict.

* * *

Some years ago sociologists of the functionalist school
formulated an explanation of the impact of industrial technology on the modern family. Although that explanation was
not empirically verified, it has become almost universally
accepted.[1] Despite some differences in emphasis, the basic
tenets of the traditional interpretation can be roughly summarized as follows:

Before industrialization the family was the basic social
unit. Most families were rural, large, and self-sustaining;
they produced and processed almost everything that was

*Reprinted by permission from Technology & Culture 17
(January, 1976), pp. 1-23. Copyright © 1976 by The University of Chicago Press.

needed for their own support and for trading in the marketplace, while at the same time performing a host of other
functions ranging from mutual protection to entertainment.
In these preindustrial families women (adult women, that is)
had a lot to do, and their time was almost entirely absorbed
by household tasks. Under industrialization the family is
much less important. The household is no longer the focus
of production; production for the marketplace and production
for sustenance have been removed to other locations. Families are smaller and they are urban rather than rural. The
number of social functions they perform is much reduced,
until almost all that remains is consumption, socialization
of small children, and tension management. As their functions diminished, families became atomized; the social bonds
that had held them together were loosened. In these post-
industrial families women have very little to do, and the
tasks with which they fill their time have lost the social
utility that they once possessed. Modern women are in
trouble, the analysis goes, because modern families are in
trouble; and modern families are in trouble because industrial technology has either eliminated or eased almost all
their former functions, but modern ideologies have not kept
pace with the change. The results of this time lag are
several: some women suffer from role anxiety, others land
in the divorce courts, some enter the labor market, and
others take to burning their brassieres and demanding liberation.

     This sociological analysis is a cultural artifact of
vast importance. Many Americans believe that it is true
and act upon that belief in various ways: some hope to re-
establish family solidarity by relearning lost productive
crafts--baking bread, tending a vegetable garden--others
dismiss the women's liberation movement as "simply a
bunch of affluent housewives who have nothing better to do
with their time." As disparate as they may seem, these
reactions have a common ideological source--the standard
sociological analysis of the impact of technological change
on family life.

     As a theory this functionalist approach has much to
recommend it, but at present we have very little evidence
to back it up. Family history is an infant discipline, and
what evidence it has produced in recent years does not lend
credence to the standard view. [2] Phillippe Ariès has shown,
for example, that in France the ideal of the small nuclear
family predates industrialization by more than a century. [3]

Historical demographers working on data from English and
French families have been surprised to find that most families were quite small and that several generations did not
ordinarily reside together; the extended family, which is
supposed to have been the rule in preindustrial societies,
did not occur in colonial New England either.[4] Rural English families routinely employed domestic servants, and even
very small English villages had their butchers and bakers
and candlestick makers; all these persons must have eased
some of the chores that would otherwise have been the housewife's burden.[5] Preindustrial housewives no doubt had much
with which to occupy their time, but we may have reason to
wonder whether there was quite as much pressure on them
as sociological orthodoxy has led us to suppose. The large
rural family that was sufficient unto itself back there on the
prairies may have been limited to the prairies--or it may
never have existed at all (except, that is, in the reveries of
sociologists).

Even if all the empirical evidence were to mesh with
the functionalist theory, the theory would still have problems,
because its logical structure is rather weak. Comparing the
average farm family in 1750 (assuming that you knew what
that family was like) with the average urban family in 1950
in order to discover the significant social changes that had
occurred is an exercise rather like comparing apples with
oranges; the differences between the fruits may have nothing
to do with the differences in their evolution. Transferring
the analogy to the case at hand, what we really need to know
is the difference, say, between an urban laboring family of
1750 and an urban laboring family 100 and then 200 years
later, or the difference between the rural nonfarm middle
classes in all three centuries, or the difference between the
urban rich yesterday and today. Surely in each of these
cases the analyses will look very different from what we
have been led to expect. As a guess we might find that for
the urban laboring families the changes have been precisely
the opposite of what the model predicted; that is, that their
family structure is much firmer today than it was in centuries past. Similarly, for the rural nonfarm middle class
the results might be equally surprising; we might find that
married women of that class rarely did any housework at all
in 1890 because they had farm girls as servants, whereas in
1950 they bore the full brunt of the work themselves. I
could go on, but the point is, I hope, clear: in order to
verify or falsify the functionalist theory, it will be necessary
to know more than we presently do about the impact of

industrialization on families of similar classes and geographical locations.

* * *

With this problem in mind I have, for the purposes of this initial study, deliberately limited myself to one kind of technological change affecting one aspect of family life in only one of the many social classes of families that might have been considered. What happened, I asked, to middle-class American women when the implements with which they did their everyday household work changed? Did the technological change in household appliances have any effect upon the structure of American households, or upon the ideologies that governed the behavior of American women, or upon the functions that families needed to perform? Middle-class American women were defined as actual or potential readers of the better-quality women's magazines, such as the Ladies' Home Journal, American Home, Parents' Magazine, Good Housekeeping, and McCall's. 6 Nonfictional material (articles and advertisements) in those magazines was used as a partial indicator of some of the technological and social changes that were occurring.

The Ladies' Home Journal has been in continuous publication since 1886. A casual survey of the nonfiction in the Journal yields the immediate impression that that decade between the end of World War I and the beginning of the depression witnessed the most drastic changes in patterns of household work. Statistical data bear out this impression. Before 1918, for example, illustrations of homes lit by gaslight could still be found in the Journal; by 1928 gaslight had disappeared. In 1917 only one-quarter (24.3 per cent) of the dwellings in the United States had been electrified, but by 1920 this figure had doubled (47.4 per cent--for rural nonfarm and urban dwellings), and by 1930 it had risen to four-fifths per cent. 7 If electrification had meant simply the change from gas or oil lamps to electric lights, the changes in the housewife's routines might not have been very great (except for eliminating the chore of cleaning and filling oil lamps); but changes in lighting were the least of the changes that electrification implied. Small electric appliances followed quickly on the heels of the electric light, and some of those augured much more profound changes in the housewife's routine.

Ironing, for example, had traditionally been one of the

most dreadful household chores, especially in warm weather when the kitchen stove had to be kept hot for the better part of the day; irons were heavy and they had to be returned to the stove frequently to be reheated. Electric irons eased a good part of this burden.[8] They were relatively inexpensive and very quickly replaced their predecessors; advertisements for electric irons first began to appear in the ladies' magazines after the war, and by the end of the decade the old flatiron had disappeared; by 1929 a survey of 100 Ford employees revealed that ninety-eight of them had the new electric irons in their homes.[9]

Data on the diffusion of electric washing machines are somewhat harder to come by; but it is clear from the advertisements in the magazines, particularly advertisements for laundry soap, that by the middle of the 1920s those machines could be found in a significant number of homes. The washing machine is depicted just about as frequently as the laundry tub by the middle of the 1920s; in 1929, forty-nine out of those 100 Ford workers had the machines in their homes. The washing machines did not drastically reduce the time that had to be spent on household laundry, as they did not go through their cycles automatically and did not spin dry; the housewife had to stand guard, stopping and starting the machine at appropriate times, adding soap, sometimes attaching the drain pipes, and putting the clothes through the wringer manually. The machines did, however, reduce a good part of the drudgery that once had been associated with washday, and this was a matter of no small consequence.[10] Soap powders appeared on the market in the early 1920s, thus eliminating the need to scrape and boil bars of laundry soap.[11] By the end of the 1920s Blue Monday must have been considerably less blue for some housewives--and probably considerably less "Monday," for with an electric iron, a washing machine, and a hot water heater, there was no reason to limit the washing to just one day of the week.

Like the routines of washing the laundry, the routines of personal hygiene must have been transformed for many households during the 1920s--the years of the bathroom mania.[12] More and more bathrooms were built in older homes, and new homes began to include them as a matter of course. Before the war most bathroom fixtures (tubs, sinks, and toilets) were made out of porcelain by hand; each bathroom was custom-made for the house in which it was installed. After the war industrialization descended upon the bathroom industry; cast iron enamelware went into mass production and fittings were standardized. In 1921 the dollar value of the production of enameled sanitary fixtures was $2.4 million, the same as it had been in 1915. By 1923, just two years later, that figure had doubled to $4.8 million; it rose again, to $5.1 million, in 1925.[13] The first recessed, double-shell cast iron enameled bathtub was put on the market in the early 1920s. A decade later the standard American bathroom had achieved its standard American form: the recessed tub, plus tiled floors and walls, brass plumbing, a single-unit toilet, an enameled sink, and a medicine chest, all set into a small room which was very often 5 feet square.[14] The bathroom evolved more quickly than any other room of the house; its standardized form was accomplished in just over a decade.

Along with bathrooms came modernized systems for heating hot water: 61 per cent of the homes in Zanesville, Ohio, had indoor plumbing with centrally heated water by 1926, and 83 per cent of the homes valued over $2,000 in Muncie, Indiana, had hot and cold running water by 1935.[15] These figures may not be typical of small American cities (or even large American cities) at those times, but they do jibe with the impression that one gets from the magazines: after 1918 references to hot water heated on the kitchen range, either for laundering or for bathing, become increasingly difficult to find.

Similarly, during the 1920s many homes were outfitted with central heating; in Muncie most of the homes of the business class had basement heating in 1924; by 1935 Federal Emergency Relief Administration data for the city indicated that only 22.4 per cent of the dwellings valued over $2,000 were still heated by a kitchen stove.[16] What all these changes meant in terms of new habits for the average housewife is somewhat hard to calculate; changes there must have been, but it is difficult to know whether those changes produced an overall saving of labor and/or time. Some chores were eliminated--hauling water, heating water on the stove, maintaining the kitchen fire--but other chores were added--most notably the chore of keeping yet another room scrupulously clean.

It is not, however, difficult to be certain about the changing habits that were associated with the new American kitchen--a kitchen from which the coal stove had disappeared. In Muncie in 1924, cooking with gas was done in two out of three homes; in 1935 only five per cent of the homes valued

over $2,000 still had coal or wood stoves for cooking.[17]
After 1918 advertisements for coal and wood stoves disappeared from the Ladies' Home Journal; stove manufacturers
purveyed only their gas, oil, or electric models. Articles
giving advice to homemakers on how to deal with the trials
and tribulations of starting, stoking, and maintaining a coal
or a wood fire also disappeared. Thus it seems a safe as-
sumption that most middle-class homes had switched to the
new method of cooking by the time the depression began.
The change in routine that was predicated on the change
from coal or wood to gas or oil was profound; aside from
the elimination of such chores as loading the fuel and removing the ashes, the new stoves were much easier to light,
maintain, and regulate (even when they did not have thermo-
stats, as the earliest models did not).[18] Kitchens were,
in addition, much easier to clean when they did not have
coal dust regularly tracked through them; one writer in the
Ladies' Home Journal estimated that kitchen cleaning was
reduced by one-half when coal stoves were eliminated.[19]

Along with new stoves came new foodstuffs and new
dietary habits. Canned foods had been on the market since
the middle of the 19th century, but they did not become an
appreciable part of the standard middle-class diet until the
1920s--if the recipes given in cookbooks and in women's
magazines are a reliable guide. By 1918 the variety of
foods available in cans had been considerably expanded from
the peas, corn, and succotash of the 19th century; an American housewife with sufficient means could have purchased almost any fruit or vegetable and quite a surprising array of
ready-made meals in a can--from Heinz's spaghetti in meat
sauce to Purity Cross's lobster à la Newburg. By the middle of the 1920s home canning was becoming a lost art.
Canning recipes were relegated to the back pages of the
women's magazines; the business-class wives of Muncie reported that, while their mothers had once spent the better
part of the summer and fall canning, they themselves rarely
put up anything, except an occasional jelly or batch of toma-
toes.[20] In part this was also due to changes in the technology of marketing food; increased use of refrigerated railroad cars during this period meant that fresh fruits and
vegetables were in the markets all year round at reasonable
prices.[21] By the early 1920s convenience foods were also
appearing on American tables: cold breakfast cereals, pan-
cake mixes, bouillon cubes, and packaged desserts could be
found. Wartime shortages accustomed Americans to eating
much lighter meals than they had previously been wont to do;

and as fewer family members were taking all their meals at home (businessmen started to eat lunch in restaurants downtown, and factories and schools began installing cafeterias), there was simply less cooking to be done, and what there was of it was easier to do.[22]

* * *

Many of the changes just described--from hand power to electric power, from coal and wood to gas and oil as fuels for cooking, from one-room heating to central heating, from pumping water to running water--are enormous technological changes. Changes of a similar dimension, either in the fundamental technology of an industry, in the diffusion of that technology, or in the routines of workers, would have long since been labeled an "industrial revolution." The change from the laundry tub to the washing machine is no less profound than the change from the hand loom to the power loom; the change from pumping water to turning on a water faucet is no less destructive of traditional habits than the change from manual to electric calculating. It seems odd to speak of an "industrial revolution" connected with housework, odd because we are talking about the technology of such homely things, and odd because we are not accustomed to thinking of housewives as a labor force or of housework as an economic commodity--but despite this oddity, I think the term is altogether appropriate.

In this case other questions come immediately to mind, questions that we do not hesitate to ask, say, about textile workers in Britain in the early 19th century, but we have never thought to ask about housewives in America in the 20th century. What happened to this particular work force when the technology of its work was revolutionized? Did structural changes occur? Were new jobs created for which new skills were required? Can we discern new ideologies that influenced the behavior of the workers?

The answer to all of these questions, surprisingly enough, seems to be yes. There were marked structural changes in the work force, changes that increased the work load and the job description of the workers that remained. New jobs were created for which new skills were required; these jobs were not physically burdensome, but they may have taken up as much time as the jobs they had replaced. New ideologies were also created, ideologies which reinforced new behavioral patterns, patterns that we might not have been led to expect if we had followed the sociologists' model to the letter. Middle-class housewives, the women who must have first felt the impact of the new household technology, were not flocking into the divorce courts or the labor market or the forums of political protest in the years immediately after the revolution in their work. What they were doing was sterilizing baby bottles, shepherding their children to dancing classes and music lessons, planning nutritious meals, shopping for new clothes, studying child psychology, and hand stitching color-coordinated curtains--all of which chores (and others like them) the standard sociological model has apparently not provided for.

The significant change in the structure of the household labor force was the disappearance of paid and unpaid servants (unmarried daughters, maiden aunts, and grandparents fall in the latter category) as household workers-- and the imposition of the entire job on the housewife herself. Leaving aside for a moment the question of which was cause and which effect (did the disappearance of the servant create a demand for the new technology, or did the new technology make the servant obsolete?), the phenomenon itself is relatively easy to document. Before World War I, when illustrators in the women's magazines depicted women doing housework, the women were very often servants. When the lady of the house was drawn, she was often the person being served, or she was supervising the serving, or she was adding an elegant finishing touch to the work. Nursemaids diapered babies, seamstresses pinned up hems, waitresses served meals, laundresses did the wash, and cooks did the cooking. By the end of the 1920s the servants had disappeared from those illustrations; all those jobs were being done by housewives--elegantly manicured and coiffed, to be sure, but housewives nonetheless (compare figs. 1 and 2).

If we are tempted to suppose that illustrations in advertisements are not a reliable indicator of structural changes of this sort, we can corroborate the changes in other ways. Apparently, the illustrators really did know whereof they drew. Statistically the number of persons throughout the country employed in household service dropped from 1,851,000 in 1910 to 1,411,000 in 1920, while the number of households enumerated in the census rose from 20.3 million to 24.4 million.[23] In Indiana the ratio of households to servants increased from 13.5/1 in 1890 to 30.5/1 in 1920, and in the country as a whole the number of paid domestic servants per 1,000 population dropped from 98.9 in 1900 to

Fig. 1. The housewife as manager. (Ladies' Home Journal, April 1918. Courtesy of Lever Brothers Co.)

Fig. 2. The housewife as laundress. (Ladies' Home
Journal, August 1928. Courtesy of Colgate-Palmolive-Peet.)

58.0 in 1920.[24] The business-class housewives of Muncie reported that they employed approximately one-half as many woman-hours of domestic service as their mothers had done.[25]

In case we are tempted to doubt these statistics (and indeed statistics about household labor are particularly un-reliable, as the labor is often transient, part-time, or simply unreported), we can turn to articles on the servant problem, the disappearance of unpaid family workers, the design of kitchens, or to architectural drawings for houses. All of this evidence reiterates the same point: qualified servants were difficult to find; their wages had risen and their num-bers fallen; houses were being designed without maid's rooms; daughters and unmarried aunts were finding jobs downtown; kitchens were being designed for housewives, not for servants.[26] The first home with a kitchen that was not an entirely separate room was designed by Frank Lloyd Wright in 1934.[27] In 1937 Emily Post invented a new char-acter for her etiquette books: Mrs. Three-in-One, the woman who is her own cook, waitress, and hostess.[28] There must have been many new Mrs. Three-in-Ones abroad in the land during the 1920s.

As the number of household assistants declined, the number of household tasks increased. The middle-class housewife was expected to demonstrate competence at several tasks that previously had not been in her purview or had not existed at all. Child care is the most obvious example. The average housewife had fewer children than her mother had had, but she was expected to do things for her children that her mother would never have dreamed of doing: to pre-pare their special infant formulas, sterilize their bottles, weigh them every day, see to it that they ate nutritionally balanced meals, keep them isolated and confined when they had even the slightest illness, consult with their teachers frequently, and chauffeur them to dancing lessons, music lessons, and evening parties.[29] There was very little Freudianism in this new attitude toward child care: mothers were not spending more time and effort on their children be-cause they feared the psychological trauma of separation, but because competent nursemaids could not be found, and the new theories of child care required constant attention from well-informed persons--persons who were willing and able to read about the latest discoveries in nutrition, in the con-trol of contagious diseases, or in the techniques of behavioral psychology. These persons simply had to be their mothers.

Consumption of economic goods provides another example of the housewife's expanded job description; like child care, the new tasks associated with consumption were not necessarily physically burdensome, but they were time consuming, and they required the acquisition of new skills. [30] Home economists and the editors of women's magazines tried to teach housewives to spend their money wisely. The present generation of housewives, it was argued, had been reared by mothers who did not ordinarily shop for things like clothing, bed linens, or towels; consequently modern housewives did not know how to shop and would have to be taught. Furthermore, their mothers had not been accustomed to the wide variety of goods that were now available in the modern marketplace; the new housewives had to be taught not just to be consumers, but to be informed consumers. [31] Several contemporary observers believed that shopping and shopping wisely were occupying increasing amounts of housewives' time. [32]

Several of these contemporary observers also believed that standards of household care changed during the decade of the 1920s. [33] The discovery of the "household germ" led to almost fetishistic concern about the cleanliness of the home. The amount and frequency of laundering probably increased, as bed linen and underwear were changed more often, children's clothes were made increasingly out of washable fabrics, and men's shirts no longer had replaceable collars and cuffs. [34] Unfortunately all these changes in standards are difficult to document, being changes in the things that people regard as so insignificant as to be unworthy of comment; the improvement in standards seems a likely possibility, but not something that can be proved.

In any event we do have various time studies which demonstrate somewhat surprisingly that housewives with conveniences were spending just as much time on household duties as were housewives without them--or, to put it another way, housework, like so many other types of work, expands to fill the time available. [35] A study comparing the time spent per week in housework by 288 farm families and 154 town families in Oregon in 1928 revealed 61 hours spent by farm wives and 63.4 hours by town wives; in 1929 a U.S. Department of Agriculture study of families in various states produced almost identical results. [36] Surely if the standard sociological model were valid, housewives in towns, where presumably the benefits of specialization and electrification were most likely to be available, should have been spending

far less time at their work than their rural sisters.   However, just after World War II economists at Bryn Mawr College reported the same phenomenon: 60. 55 hours spent by
farm housewives, 78. 35 hours by women in small cities,
80. 57 hours by women in large ones--precisely the reverse
of the results that were expected. [37]   A recent survey of
time studies conducted between 1920 and 1970 concludes that
the time spent on housework by nonemployed housewives has
remained remarkably constant throughout the period. [38]   All
these results point in the same direction: mechanization of
the household meant that time expended on some jobs decreased, but also that new jobs were substituted, and in
some cases--notably laundering--time expenditures for old
jobs increased because of higher standards.   The advantages
of mechanization may be somewhat more dubious than they
seem at first glance.

* * *

As the job of the housewife changed, the connected
ideologies also changed; there was a clearly perceptible difference in the attitudes that women brought to housework
before and after World War I. [39]   Before the war the trials
of doing housework in a servantless home were discussed
and they were regarded as just that--trials, necessary
chores that had to be got through until a qualified servant
could be found.   After the war, housework changed: it was
no longer a trial and a chore, but something quite different
--an emotional "trip."   Laundering was not just laundering,
but an expression of love; the housewife who truly loved her
family would protect them from the embarrassment of tattle-
tale gray.   Feeding the family was not just feeding the
family, but a way to express the housewife's artistic incli-
nations and a way to encourage feelings of family loyalty
and affection.   Diapering the baby was not just diapering,
but a time to build the baby's sense of security and love for
the mother.   Cleaning the bathroom sink was not just cleaning, but an exercise of protective maternal instincts, providing a way for the housewife to keep her family safe from
disease.   Tasks of this emotional magnitude could not possibly be delegated to servants, even assuming that qualified
servants could be found.

Women who failed at these new household tasks were
bound to feel guilt about their failure.   If I had to choose
one word to characterize the temper of the women's magazines during the 1920s, it would be "guilt."   Readers of the

better-quality women's magazines are portrayed as feeling
guilty a good lot of the time, and when they are not guilty
they are embarrassed: guilty if their infants have not gained
enough weight, embarrassed if their drains are clogged,
guilty if their children go to school in soiled clothes, guilty
if all the germs behind the bathroom sink are not eradicated,
guilty if they fail to notice the first signs of an oncoming
cold, embarrassed if accused of having body odor, guilty if
their sons go to school without good breakfasts, guilty if
their daughters are unpopular because of old-fashioned, or
unironed, or--heaven forbid--dirty dresses (see figs. 3 and
4). In earlier times women were made to feel guilty if they
abandoned their children or were too free with their affec-
tions. In the years after World War I, American women
were made to feel guilty about sending their children to
school in scuffed shoes. Between the two kinds of guilt
there is a world of difference.

* * *

Let us return for a moment to the sociological model
with which this essay began. The model predicts that changing patterns of household work will be correlated with at
least two striking indicators of social change: the divorce
rate and the rate of married women's labor force participation. That correlation may indeed exist, but it certainly is
not reflected in the women's magazines of the 1920s and
1930s: divorce and full-time paid employment were not part
of the life-style or the life pattern of the middle-class housewife as she was idealized in her magazines.

There were social changes attendant upon the introduction of modern technology into the home, but they were
not the changes that the traditional functionalist model predicts; on this point a close analysis of the statistical data
corroborates the impression conveyed in the magazines. The
divorce rate was indeed rising during the years between the
wars, but it was not rising nearly so fast for the middle and
upper classes (who had, presumably, easier access to the
new technology) as it was for the lower classes. By almost
every gauge of socioeconomic status--income, prestige of
husband's work, education--the divorce rate is higher for
persons lower on the socioeconomic scale--and this is a
phenomenon that has been constant over time.[40]

The supposed connection between improved household
technology and married women's labor force participation

Fig. 3. Sources of housewifely guilt: the good mother smells sweet. (Ladies' Home Journal, August 1928. Courtesy of Warner-Lambert, Inc.)

Fig. 4. Sources of housewifely guilt: the good mother must be beautiful. (Ladies' Home Journal, July 1928. Courtesy of Colgate-Palmolive-Peet.)

seems just as dubious, and on the same grounds. The single socioeconomic factor which correlates most strongly (in cross-sectional studies) with married women's employment is husband's income, and the correlation is strongly negative; the higher his income, the less likely it will be that she is working.[41] Women's labor force participation increased during the 1920s but this increase was due to the influx of single women into the force. Married women's participation increased slightly during those years, but that increase was largely in factory labor--precisely the kind of work that middle-class women (who were again, much more likely to have labor-saving devices at home) were least likely to do.[42] If there were a necessary connection between the improvement of household technology and either of these two social indicators, we would expect the data to be precisely the reverse of what in fact has occurred: women in the higher social classes should have fewer functions at home and should therefore be more (rather than less) likely to seek paid employment or divorce.

Thus for middle-class American housewives between the wars, the social changes that we can document are not the social changes that the functionalist model predicts; rather than changes in divorce or patterns of paid employment, we find changes in the structure of the work force, in its skills, and in its ideology. These social changes were concomitant with a series of technological changes in the equipment that was used to do the work. What is the relationship between these two series of phenomena? Is it possible to demonstrate causality or the direction of that causality? Was the decline in the number of households employing servants a cause or an effect of the mechanization of those households? Both are, after all, equally possible. The declining supply of household servants, as well as their rising wages, may have stimulated a demand for new appliances at the same time that the acquisition of new appliances may have made householders less inclined to employ the laborers who were on the market. Are there any techniques available to the historian to help us answer these questions?

* * *

In order to establish causality, we need to find a connecting link between the two sets of phenomena, a mechanism that, in real life, could have made the causality work. In this case a connecting link, an intervening agent between the social and the technological changes, comes immediately to

mind: the advertiser--by which term I mean a combination
of the manufacturer of the new goods, the advertising agent
who promoted the goods, and the periodical that published
the promotion. All the new devices and new foodstuffs that
were being offered to American households were being manufactured and marketed by large companies which had considerable amounts of capital invested in their production:
General Electric, Procter & Gamble, General Foods, Lever
Brothers, Frigidaire, Campbell's, Del Monte, American Can,
Atlantic & Pacific Tea--these were all well-established firms
by the time the household revolution began, and they were all
in a position to pay for national advertising campaigns to
promote their new products and services. And pay they did;
one reason for the expanding size and number of women's
magazines in the 1920s was, no doubt, the expansion in
revenues from available advertisers. [43]

Those national advertising campaigns were likely to
have been powerful stimulators of the social changes that
occurred in the household labor force; the advertisers probably did not initiate the changes, but they certainly encouraged
them. Most of the advertising campaigns manifestly worked,
so they must have touched upon areas of real concern for
American housewives. Appliance ads specifically suggested
that the acquisition of one gadget or another would make it
possible to fire the maid, spend more time with the children,
or have the afternoon free for shopping. [44] Similarly, many
advertisements played upon the embarrassment and guilt
which were now associated with household work. Ralston,
Cream of Wheat, and Ovaltine were not themselves responsible for the compulsive practice of weighing infants and children repeatedly (after every meal for newborns, every day
in infancy, every week later on), but the manufacturers certainly did not stint on capitalizing upon the guilt that women
apparently felt if their offspring did not gain the required
amounts of weight. [45] And yet again, many of the earliest
attempts to spread "wise" consumer practices were under-
taken by large corporations and the magazines that desired
their advertising: mail-order shopping guides, "product-
testing" services, pseudoinformative pamphlets, and other
such promotional devices were all techniques for urging the
housewife to buy new things under the guise of training her
in her role as skilled consumer. [46]

Thus the advertisers could well be called the "ideologues" of the 1920s, encouraging certain very specific social
changes--as ideologues are wont to do. Not surprisingly, the

changes that occurred were precisely the ones that would
gladden the hearts and fatten the purses of the advertisers;
fewer household servants meant a greater demand for labor
and timesaving devices; more household tasks for women
meant more and more specialized products that they would
need to buy; more guilt and embarrassment about their failure
to succeed at their work meant a greater likelihood that they
would buy the products that were intended to minimize that
failure.   Happy, full-time housewives in intact families spend
a lot of money to maintain their households; divorced women
and working women do not.   The advertisers may not have
created the image of the ideal American housewife that dominated the 1920s--the woman who cheerfully and skillfully set
about making everyone in her family perfectly happy and
perfectly healthy--but they certainly helped to perpetuate it.

The role of the advertiser as connecting link between
social change and technological change is at this juncture
simply a hypothesis, with nothing much more to recommend
it than an argument from plausibility.   Further research may
serve to test the hypothesis, but testing it may not settle
the question of which was cause and which effect--if that
question can ever be settled definitively in historical work.
What seems most likely in this case, as in so many others,
is that cause and effect are not separable, that there is a
dynamic interaction between the social changes that married
women were experiencing and the technological changes that
were occurring in their homes.   Viewed this way, the disappearance of competent servants becomes one of the factors
that stimulated the mechanization of homes, and this mechanization of homes becomes a factor (though by no means the
only one) in the disappearance of servants.   Similarly, the
emotionalization of housework becomes both cause and effect
of the mechanization of that work; and the expansion of time
spent on new tasks becomes both cause and effect of the introduction of time-saving devices.   For example the social pressure to spend more time in child care may have led to a decision to purchase the devices; once purchased, the devices
could indeed have been used to save time--although often they
were not.

* * *

If one holds the question of causality in abeyance, the
example of household work still has some useful lessons to
teach about the general problem of technology and social
change.   The standard sociological model for the impact of

modern technology on family life clearly needs some revision: at least for middle-class nonrural American families in the 20th century, the social changes were not the ones that the standard model predicts. In these families the functions of at least one member, the housewife, have increased rather than decreased; and the dissolution of family life has not in fact occurred.

Our standard notions about what happens to a work force under the pressure of technological change may also need revision. When industries become mechanized and rationalized, we expect certain general changes in the work force to occur: its structure becomes more highly differentiated, individual workers become more specialized, managerial functions increase, and the emotional context of the work disappears. On all four counts our expectations are reversed with regard to household work. The work force became less rather than more differentiated as domestic servants, unmarried daughters, maiden aunts, and grandparents left the household and as chores which had once been performed by commercial agencies (laundries, delivery services, milkmen) were delegated to the housewife. The individual workers also became less specialized; the new housewife was now responsible for every aspect of life in her household, from scrubbing the bathroom floor to keeping abreast of the latest literature in child psychology.

The housewife is just about the only unspecialized worker left in America--a veritable jane-of-all-trades at a time when the jacks-of-all-trades have disappeared. As her work became generalized the housewife was also proletarianized: formerly she was ideally the manager of several other subordinate workers; now she was idealized as the manager and the worker combined. Her managerial functions have not entirely disappeared, but they have certainly diminished and have been replaced by simple manual labor; the middle-class, fairly well educated housewife ceased to be a personnel manager and became, instead, a chauffeur, chairwoman, and short-order cook. The implications of this phenomenon, the proletarianization of a work force that had previously seen itself as predominantly managerial, deserve to be explored at greater length than is possible here, because I suspect that they will explain certain aspects of the women's liberation movement of the 1960s and 1970s which have previously eluded explanation: why, for example, the movement's greatest strength lies in social and economic groups who seem, on the surface at least, to need it least-- women who are white, well-educated, and middle-class.

Finally, instead of desensitizing the emotions that were connected with household work, the industrial revolution in the home seems to have heightened the emotional context of the work, until a woman's sense of self-worth became a function of her success at arranging bits of fruit to form a clown's face in a gelatin salad. That pervasive social illness, which Betty Friedan characterized as "the problem that has no name," arose not among workers who found that their labor brought no emotional satisfaction, but among workers who found that their work was invested with emotional weight far out of proportion to its own inherent value: "How long," a friend of mind is fond of asking, "can we continue to believe that we will have orgasms while waxing the kitchen floor?"

Notes

1. For some classic statements of the standard view, see W. F. Ogburn and M. F. Nimkoff, Technology and the Changing Family (Cambridge, Mass., 1955); Robert F. Winch, The Modern Family (New York, 1952); and William J. Goode, The Family (Englewood Cliffs, N.J., 1964).

2. This point is made by Peter Laslett in "The Comparative History of Household and Family," in The American Family in Social Historical Perspective, ed. Michael Gordon (New York, 1973), pp. 28-29.

3. Phillippe Ariès, Centuries of Childhood: A Social History of Family Life (New York, 1960).

4. See Laslett, pp. 20-24; and Philip J. Greven, "Family Structure in Seventeenth Century Andover, Massachusetts," William and Mary Quarterly 23 (1966); 234-56.

5. Peter Laslett, The World We Have Lost (New York, 1965), passim.

6. For purposes of historical inquiry, this definition of middle-class status corresponds to a sociological reality, although it is not, admittedly, very rigorous. Our contemporary experience confirms that there are class differences reflected in magazines, and this situation seems to have existed in the past

as well. On this issue see Robert S. Lynd and
Helen M. Lynd, Middletown: A Study in Contemporary American Culture (New York, 1929), pp. 240-
44, where the marked difference in magazines sub-
scribed to by the business-class wives as opposed
to the working-class wives is discussed; Salme
Steinberg, "Reformer in the Marketplace: E. W.
Bok and The Ladies' Home Journal" (Ph. D. diss.,
Johns Hopkins University, 1973), where the conscious
attempt of the publisher to attract a middle-class
audience is discussed; and Lee Rainwater et al.,
Workingman's Wife (New York, 1959), which was
commissioned by the publisher of working-class
women's magazines in an attempt to understand the
attitudinal differences between working-class and
middle-class women.

7. Historical Statistics of the United States, Colonial Times
to 1957 (Washington, D.C., 1960), p. 510.

8. The gas iron, which was available to women whose
homes were supplied with natural gas, was an
earlier improvement on the old-fashioned flatiron,
but this kind of iron is so rarely mentioned in the
sources that I used for this survey that I am unable
to determine the extent of its diffusion.

9. Hazel Kyrk, Economic Problems of the Family (New
York, 1933), p. 368, reporting a study in Monthly
Labor Review 30 (1930): 1909-52.

10. Although this point seems intuitively obvious, there is
some evidence that it may not be true. Studies of
energy expenditure during housework have indicated
that by far the greatest effort is expended in hauling
and lifting the wet wash, tasks which were not
eliminated by the introduction of washing machines.
In addition, if the introduction of the machines
served to increase the total amount of wash that
was done by the housewife, this would tend to cancel the energy-saving effects of the machines themselves.

11. Rinso was the first granulated soap; it came on the
market in 1918. Lux Flakes had been available
since 1906; however it was not intended to be a
general laundry product but rather one for laundering

delicate fabrics. "Lever Brothers," *Fortune* 26 (November 1940): 95.

12. I take this account, and the term, from Lynd and Lynd, p. 97. Obviously, there were many American homes that had bathrooms before the 1920s, particularly urban row houses, and I have found no way of determining whether the increases of the 1920s were more marked than in previous decades. The rural situation was quite different from the urban; the President's Conference on Home Building and Home Ownership reported that in the late 1920s, 71 per cent of the urban families surveyed had bathrooms, but only 33 per cent of the rural families did (John M. Gries and James Ford, eds., *Homemaking, Home Furnishing and Information Services*, President's Conference on Home Building and Home Ownership, vol. 10 [Washington, D.C., 1932], p. 13).

13. The data above come from Siegfried Giedion, *Mechanization Takes Command* (New York, 1948), pp. 685-703.

14. For a description of the standard bathroom see Helen Sprackling, "The Modern Bathroom," *Parents' Magazine* 8 (February 1933): 25.

15. *Zanesville, Ohio and Thirty-six Other American Cities* (New York, 1927), p. 65. Also see Robert S. Lynd and Helen M. Lynd, *Middletown in Transition* (New York, 1936), p. 537. Middletown is Muncie, Indiana.

16. Lynd and Lynd, *Middletown*, p. 96, and *Middletown in Transition*, p. 539.

17. Lynd and Lynd, *Middletown*, p. 98, and *Middletown in Transition*, p. 562.

18. On the advantages of the new stoves, see *Boston Cooking School Cookbook* (Boston, 1916), pp. 15-20; and Russell Lynes, *The Domesticated Americans* (New York, 1957), pp. 119-20.

19. "How to Save Coal While Cooking," *Ladies' Home Journal* 25 (January 1908): 44.

20. Lynd and Lynd, *Middletown*, p. 156.

21. Ibid.; see also "Safeway Stores," Fortune 26 (October 1940): 60.

22. Lynd and Lynd, Middletown, pp. 134-35 and 153-54.

23. Historical Statistics, pp. 16 and 77.

24. For Indiana data, see Lynd and Lynd, Middletown, p. 169. For national data, see D. L. Kaplan and M. Claire Casey, Occupational Trends in the United States, 1900-1950, U.S. Bureau of the Census Working Paper no. 5 (Washington, D.C., 1958), table 6. The extreme drop in numbers of servants between 1910 and 1920 also lends credence to the notion that this demographic factor stimulated the industrial revolution in housework.

25. Lynd and Lynd, Middletown, p. 169.

26. On the disappearance of maiden aunts, unmarried daughters, and grandparents, see Lynd and Lynd, Middletown, pp. 25, 99, and 110; Edward Bok, "Editorial," American Home 1 (October 1928): 15; "How to Buy Life Insurance," Ladies' Home Journal 45 (March 1928): 35. The house plans appeared every month in American Home, which began publication in 1928. On kitchen design, see Giedion, pp. 603-21; "Editorial," Ladies' Home Journal 45 (April 1928): 36; advertisement for Hoosier kitchen cabinets, Ladies' Home Journal 45 (April 1928): 117. Articles on servant problems include "The Vanishing Servant Girl," Ladies Home Journal 35 (May 1918): 48; "Housework, Then and Now," American Home 8 (June 1932): 128; "The Servant Problem," Fortune 24 (March 1938): 80-84; and Report of the YWCA Commission on Domestic Service (Los Angeles, 1915).

27. Giedion, p. 619. Wright's new kitchen was installed in the Malcolm Willey House, Minneapolis.

28. Emily Post, Etiquette: The Blue Book of Social Usage, 5th ed. rev. (New York, 1937), p. 823.

29. This analysis is based upon various child-care articles that appeared during the period in the Ladies' Home Journal, American Home, and Parents' Magazine. See also Lynd and Lynd, Middletown, chapt. 11.

30. John Kenneth Galbraith has remarked upon the advent of woman as consumer in Economics and the Public Purpose (Boston, 1973), pp. 29-37.

31. There was a sharp reduction in the number of patterns for home sewing offered by the women's magazines during the 1920s; the patterns were replaced by articles on "what is available in the shops this season." On consumer education see, for example, "How to Buy Towels," Ladies' Home Journal 45 (February 1928): 134; "Buying Table Linen," Ladies' Home Journal 45 (March 1928): 43; and "When the Bride Goes Shopping," American Home 1 (January 1928): 370.

32. See, for example, Lynd and Lynd, Middletown, pp. 176 and 196; and Margaret G. Reid, Economics of Household Production (New York, 1934), chap. 13.

33. See Reid, pp. 64-68; and Kyrk, p. 98.

34. See advertisement for Cleanliness Institute--"Self-respect thrives on soap and water," Ladies' Home Journal 45 (February 1928): 107. On changing bed linen, see "When the Bride Goes Shopping," American Home 1 (January 1928): 370. On laundering children's clothes, see, "Making a Layette," Ladies' Home Journal 45 (January 1928): 20; and Josephine Baker, "The Youngest Generation," Ladies' Home Journal 45 (March 1928); 185.

35. This point is also discussed at length in my paper "What Did Labor-saving Devices Really Save?" (unpublished).

36. As reported in Kyrk, p. 51.

37. Bryn Mawr College Department of Social Economy, Women During the War and After (Philadelphia, 1945); and Ethel Goldwater, "Woman's Place," Commentary 4 (December 1947): 578-85.

38. JoAnn Vanek, "Keeping Busy: Time Spent in Housework, United States, 1920-1970" (Ph.D. diss., University of Michigan, 1973). Vanek reports an average of 53 hours per week over the whole period. This figure is significantly lower than the figures

reported above, because each time study of housework has been done on a different basis, including
different activities under the aegis of housework,
and using different methods of reporting time expenditures; the Bryn Mawr and Oregon studies are
useful for the comparative figures that they report
internally, but they cannot easily be compared with
each other.

39. This analysis is based upon my reading of the middle-
class women's magazines between 1918 and 1930.
For detailed documentation see my paper "Two
Washes in the Morning and a Bridge Party at Night:
The American Housewife Between the Wars," Wom-
en's Studies (in press) [now published in Women's
Studies, III (1976):147-72]. It is possible that the appearance of guilt as a strong element in advertising
is more the result of new techniques developed by the
advertising industry than the result of attitudinal
changes in the audience--a possibility that I had not
considered when doing the initial research for this
paper. See A. Michael McMahon, "An American
Courtship: Psychologists and Advertising Theory in
the Progressive Era," American Studies 13 (1972):
5-18.

40. For a summary of the literature on differential divorce
rates, see Winch, p. 706; and William J. Goode,
After Divorce (New York, 1956) p. 44. The earliest
papers demonstrating this differential rate appeared
in 1927, 1935, and 1939.

41. For a summary of the literature on married women's
labor force participation, see Juanita Kreps, Sex
in the Marketplace: American Women at Work
(Baltimore, 1971), pp. 19-24.

42. Valerie Kincaid Oppenheimer, The Female Labor Force
in the United States, Population Monograph Series,
no. 5 (Berkeley, 1970), pp. 1-15; and Lynd and
Lynd, Middletown, pp. 124-27.

43. On the expanding size, number, and influence of
women's magazines during the 1920s, see Lynd
and Lynd, Middletown, pp. 150 and 240-44.

44. See, for example, the advertising campaigns of General
Electric and Hotpoint from 1918 through the rest of

the 1920s; both campaigns stressed the likelihood
that electric appliances would become a thrifty replacement for domestic servants.

45. The practice of carefully observing children's weight
was initiated by medical authorities, national and
local governments, and social welfare agencies, as
part of the campaign to improve child health which
began about the time of World War I.

46. These practices were ubiquitous, American Home, for
example, which was published by Doubleday, assisted its advertisers by publishing a list of in-
formative pamphlets that readers could obtain; devoting half a page to an index of its advertisers;
specifically naming manufacturer's and list prices
in articles about products and services; allotting
almost one-quarter of the magazine to a mail-order
shopping guide which was not (at least ostensibly)
paid advertisement; and as part of its editorial
policy, urging its readers to buy new goods.

INTRODUCTION TO PART II(B)
AND GENERAL CONCLUSION

Females as Children and as Bearers
and Rearers of Children

Just as Ruth Cowan has noted the lag between the
development of household technologies and societal values
which encourage men to assume a major role in housework,
Vern L. Bullough similarly discusses the "cultural lag" between the availability of three technologies concerned with
female physiology and society's apparent willingness to en-
gage in role-reversal of the sexes in the domestic spheres
in this century. Bullough views the rise of contraceptives,
nursing technologies and the sanitary pad, and shows that
society has long had at its disposal practicable devices which
could reduce women's confinement to the home and to child-
raising. Undoubtedly, he argues, such technologies have
ultimately made possible women's greater mobility in society and in the economy, but social values encouraging men
as child-raisers have not developed rapidly. (It is indeed
interesting to note the falling birth rate as women increasingly move into careers and jobs outside the home. This
indicates, for one thing, that rather than males' devoting
much more time to child-raising, couples choose simply to
have fewer children. It is also instructive to realize that
women have typically borne the brunt of risk in contraception in modern society, as the "pill," the diaphragm and
other sometimes harmful contraceptives have been developed
mainly for women. Also, unsafe abortions have threatened,
harmed and killed women for centuries.)

Also dealing with technology and child-raising, Carroll W. Pursell, Jr. has examined the socialization of children into traditional sex roles through the medium of toys.
During the 1920s and 1930s, little girls were consistently
educated for the roles of housewife and mother by dolls and

toys which simulated various household technologies. Little
boys, on the other hand, were encouraged to be scientists
and engineers, among other professions, with chemistry
sets, Erector sets, various vehicles and tools. Indeed,
Martha Trescott has preliminarily found in her investigation
of the history of women engineers in the U.S. that the effects of such socialization on girls truly have helped to account for the persistently low numbers of women who choose
careers in science and technology. Pressure from peers
and parents, along with school authorities, has been significant here, and the toys with which parents allow children to
play are a part of this entire picture.

According to the work of Cowan, Bullough and Pursell, one might indeed conclude that the twentieth century
has represented very little "progress" in role-reversal
among the sexes up until the current decade, if then. Indeed, the women's movement of the 1960s and 1970s was
undoubtedly, in part, a response to this severe and strongly
entrenched sex-role stereotyping of women, which in many
ways seems to have been underscored in this century. And
this, despite the many important technological changes which
might have been expected to effect more sex-role reversal.
Implicitly, both Pursell and Cowan treat one technology (or
system of many technologies) which has conduced to this
sex-typing: the media. In future work concerning women
and technological change, this technology and its innovators
and practitioners must be viewed for their part in the im-
becilization of women in modern times. Many of the leaders
of the current women's movement, such as Betty Friedan,
have faulted the media here. Historians can ill afford not
to use their own medium (history) to analyze this area of
technology and industry when considering relationships between technologies and social values affecting sexual functions and roles.

As stated in the general introduction, the essays in
this book tend to point out that basic sex roles have remained quite stable in the face of great technological changes
in our culture. In fact, one conclusion of the book might
be that, in general, social values overwhelmingly determine
the course and use of technology, rather than vice-versa.
Indeed, we must generally look to other forces besides a
mythical technological monolith to determine why society has
evolved as it has.

It is worthwhile to bear in mind that technology is

not an impersonal force, one function of which has been the oppression of women. But at the same time, our technological systems have been almost totally dominated by male decision-makers and designers. What this has to tell us about why our technologies have not more speedily liberated women from housework, child-raising, unequal pay for equal (and, in some cases, superior) work and dead-end jobs cannot here be investigated. Surely, many more analyses of sex roles, sex-typing of work and technological change are needed. It is hoped that this volume has served to introduce this field of inquiry and at least to place some of the questions about women and technological change in proper perspective.

1. FEMALE PHYSIOLOGY, TECHNOLOGY AND WOMEN'S LIBERATION

by Vern L. Bullough

I

The impact of technology upon women and their role in contemporary society has been the subject of much theorizing but little detailed historical research. Sociologists such as W. F. Ogburn, M. F. Nimkoff, Robert S. Winch, and William J. Goode have in general argued that industrial society has led to smaller families, a decrease in the tasks performed by women in their home, and reduction of their social functions until all that remains is "consumption, socialization of small children, and tension management."[1] Ruth Schwartz Cowan summarized the generally accepted view:

> In these post-industrial families women have very little to do, and the tasks with which they fill their time have lost the social utility that they once possessed. Modern women are in trouble, the analysis goes, because ... industrial technology has either eliminated or eased almost all their former functions, but modern ideologies have not kept pace with the change. The results of this time lag are several: some women suffer from role anxiety, others land in the divorce courts, some enter the labor market, and others take to burning their brassieres and demanding liberation.[2]

After examining the effects of the "industrial revolution" in the home, Professor Cowan disagreed with this view. She argued that though there were social changes stemming from the application of modern technology to housework, the changes were not necessarily the ones the traditional func-

236

tionalist model would have predicted. Instead, she found
that the tasks of the middle-class housewife tended to increase rather than decrease. Where the upper income-level
housewife earlier had been regarded as the manager of several subordinates, the effect of technological innovation in
the home had been to make her more of a worker, a woman
of all trades. This, Cowan argued, led to a proletarianization of the work force that previously had seen itself as
predominantly managerial. In her opinion, the women's
liberation movement of the 1960s and 1970s was, in part at
least, motivated by recognition of this status loss.[3]

There is still another aspect of the technological picture that needs to be examined, one that not only supplements Professor Cowan's explanation but also reinforces
some of the traditional functionalist sociological explanations
about the effect of change in woman's role. In particular,
this paper will focus on technological innovations which lessened the physiological handicaps of being female in a male-
dominated world. These handicaps include weaker muscular
structure and what has been called the "female secretions"
derived from the fact that women menstruate, ovulate and
bear children and lactate. While the female of the species
will continue to bear the children, the more confining aspects
of being female have been largely overcome by technology.
In the process, much of the biological justification for the
stereotyped role model has been eliminated, and the groundwork laid for challenging traditional patterns of marriage
and family life.

In this paper the technological innovations which have
lessened the muscular advantage of males over females have
been ignored, since in the long run these have benefited men
as much as women. It should, however, be pointed out that
females proportionately have smaller bones, fewer muscle
cells, and lay down fat differently from the males. Female
hormones also work differently from male ones; estrogen,
for example, is implicated in the decelerating growth curve
which terminates in the union of the epiphyses, thus termi-
nating bone growth. This, in the past, put women at a disadvantage to men in those tasks requiring strong muscular
effort. (The fact that female athletes may have to undergo
chromosome tests indicates that this particular physiological
advantage of the male over the female is still believed to
exist, even though technology has lessened its importance.)

Still, the more important challenges to the traditional

role models for women have come from technological developments associated with pregnancy and child-rearing--namely, contraceptives, the infant-feeding bottle, pasteurization of milk, and the manufacture of the sanitary pad, the subjects of this paper. These technological innovations were allied with medical breakthroughs which made it possible to control anemia, cystitis, and other physiological vulnerabilities associated with being female.

II

First, and perhaps the most important, is the development of contraceptives. This development was closely associated with the vulcanization of rubber in the middle of the nineteenth century. Under the most favorable conditions it has been estimated that an average woman could become pregnant every other year during her reproductive life (normally between 15 and 45 years of age). This would mean that theoretically a woman could give birth to about fifteen children, although the actual recorded rate of a society has hardly ever approached as high as ten, even among the most prolific groups. (In the United States today one group, the Hutterites, has a maternity ratio of 10.6, that is, the average number of live births per woman aged 45 or over, regardless of whether a particular woman has children.)[4] The possibility of pregnancy undoubtedly tended to be used to restrict the physical freedom of women in the past, while actual pregnancies, at least in any great numbers, tended to weaken a woman physically. Though there are various ways by which the maternity ratio historically has been lessened, such as abstinence, regulating marriage age, prostitution, coitus interruptus, et al., most such methods tended to perpetuate a double sexual standard which carried over into other aspects of woman's role.[5] It would seem logical, then, that woman's ability to control whether or not she would become pregnant would be a major factor in encouraging role change and in giving greater freedom, in weakening the grounds for the double standard.

We know that there was a decline in fertility in some of the more developed countries in the last part of the nineteenth and the first part of the twentieth century, and much research on this has been carried out at the Office of Population Research at Princeton University during the last few decades.[6] Various factors were involved in this decline, including changing expectations about standard of living and rising ambitions of parents for their children, [7] but the most important factor was the availability of contraceptives.

Prior to the last part of the nineteenth century, the standard mechanical contraceptives were the condom and the sponge, supplemented by various douching substances. The condom, made either of animal caeces or fish bladders, was primarily used as a prophylactic against disease. Many users, as well as their female partners, found them uncomfortable and a hindrance to sexual pleasure. The biggest obstacle to their use was their expense. The sponge might or might not have been effective depending on the solutions it had been soaked in before placement in the vagina. Douching in general was not very effective, [8] although even a partially effective contraceptive would affect the maternity ratio. Before a contraceptive could begin to give women some feeling of freedom, it would have to approach 80 to 90 per cent effectiveness. This rate only began to appear towards the end of the nineteenth century, in large part because of the new rubber products made possible by the vulcanization of rubber in 1843-1844 by Charles Goodyear and Thomas Hancock.

Even before rubber had been vulcanized, hard rubber rings had been used as a pessary, a device inserted into the vagina to support the uterus or rectum and used by women who had prolapses of the uterus. The contraceptive value of such devices was not ignored and as early as 1838 a German, Friedrich Adolph Wilde, urged their use for this purpose. Wilde himself made a pessary of elastic resinous material formed from a wax impression of the cervix and designed to cover the os; he later used rubber wax as his material. [9] The association of the pessary with both a medical and a contraceptive purpose is important since contraceptives per se often ran into legal difficulties. Pessaries never did. The diaphragm called the "Womb Veil," which Edward Bliss Foote distributed in the United States after 1860, [10] is a case in point. Both it and the pamphlets describing its use were seized and destroyed under the so-called Comstock Law of 1873. Patents, however, were permissible under the title "pessary" since the pessary had other implications than simply contraception, including offering support for a prolapsed uterus.

The most widely used of these contraceptive devices was the Mensinga pessary developed by W. P. J. Mensinga in 1842. Mensinga had taken an ordinary hard rubber ring, similar to those used by physicians for correcting displacements and prolapse of the uterus, and covered it with sheet rubber to form a diaphragm across the vagina. With the vulcanization of rubber, Mensinga improved his pessary, and, adding to its effectiveness, incorporated a flat watchspring in the edge which helped keep it in place. Later a coiled spring was used.[11] The use of the Mensinga cap spread rapidly in the last part of the nineteenth century, particularly after 1882 when Aletta Jacobs, a student of Mensinga, opened the first modern birth control clinic in the Netherlands. She used the Mensinga for her patients. The first mention in English of the Mensinga cap was in 1887 in a book by a Leeds physician, The Wife's Handbook, and for this he was expelled from the General Medical Council.[12] In the United States the Mensinga cap was not mentioned in any formal medical way until 1923, when James T. Cooper gave a paper summarizing the results of fifty users at the Midwestern Birth Control Conference in Chicago.[13]

Invention obviously does not necessarily imply use, and undoubtedly the usage of the Mensinga diaphragm was not particularly widespread in the nineteenth-century United States. Even in the twentieth century Margaret Sanger allegedly had difficulty in finding out information about contraceptives. Still, such devices likely were ubiquitous. Some evidence of widespread diffusion is reflected by the numerous patent applications for pessaries, a necessary euphemism since the U. S. Patent Office refused to patent contraceptive devices in the nineteenth century.

Rhodes Lockwood of Boston, for example, filed a pessary request on July 30, 1877, very similar to the Mensinga one. He requested a patent (which was granted) for a pessary made of a metallic spring core, lapped, reinforced by a strengthening strip and covered with vulcanized rubber soft enough to be easily bent or collapsed when applying it, and soft enough to yield as a cushion and not distress the wearer. In his patent application he clearly indicates the advantages of the vulcanized rubber pessary:

> A pessary of hard rubber is very stiff and liable to be broken, is uncomfortable to the wearer, and cannot be readily fitted to any patient.

> A pessary covered with gutta-percha or unvulcanized india-rubber quickly collects sediment, and emits an

unpleasant odor, from which the soft vulcanized
cover is perfectly free.

This improved pessary may be bent or folded in
any way necessary to apply it, and it will then
spring outward under the action of its spring.
This vulcanized soft-rubber covering, free from
smell gained in use, and soft, so as to act as a
cushion, and not inflame or feel harsh or chafe
the parts, is a matter of great importance.[14]

Several other patents were issued in the 1870s for
pessaries of one type or another, some being more clearly
designed to deal with a prolapsed uterus while others were
obviously conceived as contraceptive devices. None of the
inventors advertised, as did Foote, that its use put concep-
tion "entirely under the control of the wife, " and without any
infringement upon the pleasure of intercourse. Obviously,
the Comstock Law more or less guaranteed that such devices
would be restricted to the middle- or upper-middle-class
woman who consulted physicians who, in turn, might diag-
nose the desire not to get pregnant as a prolapsed uterus.

Equally important as the pessary was the condom,
although it had the disadvantage of being controlled by the
male rather than the female. The history of the rubber
condom, however, is more difficult to trace than the diaphragm's, simply because it was more obviously a contraceptive device, although it too was justified as a prophylactic. Rubber technology was more important in the development of the rubber condom than the diaphragm, if only
because the condom required more sophisticated methods.
These techniques, however, had been patented by the middle
of the 1850s. In 1848 Charles Goodyear received a U.S.
Patent (5536) for manufacture of hollow rubber articles,
and in that same year Hancock was granted a British patent
(12153) for glass moulds to be used with gutta-percha. The
sulphurous dioxide method of treating rubber had been patented by Alexander Parkes in 1846 and liquid latex had been
developed in 1853. When did rubber condoms appear? The
symbolic date for their first introduction is 1876, and the
place, the World Exposition held in Philadelphia to celebrate
America's Centennial.[16] They must have appeared much
earlier, however, and the earliest description of a rubber
apex envelope designed to cover the glans of the penis is
1862. Full length condoms appear by 1869, and their use
was widely recommended in an influential text in 1872.[17]

The earliest condoms were moulded from sheet crepe, and although a satisfactory standard of vulcanization could be achieved, the finished product carried a seam along its entire length. This method was soon replaced by the seamless cement process, so named because the process was similar to that used in producing rubber cement. The natural rubber was ground up, dissolved and heated with solvents into which cylindrical glass moulds were dipped. As the solvent evaporated, the condoms dried, ready for vulcanization, which was accomplished by exposure to sulphur dioxide. The process was later used in making rubber gloves which appeared on the market about 1889,[18] evidently several years after the process had been used in manufacturing condoms. The process was time-consuming as well as hazardous to workers, since the solvents were highly inflammable. The major defect of the finished product was that it deteriorated rather rapidly.

The sale and distribution of contraceptives is a subject beyond the scope of this paper, although obviously important to the hypothesis stated here. We know that in England barber shops and various emporia which had established mail order trade in books and pamphlets, the predecessor of the early twentieth-century "surgical stores" and the mid-twentieth century "eros" stores, were important in this respect. On the continent there were official contraceptive clinics. In the United States the ubiquitous local drug store seems to have been the key. One of the earliest references to the sale of condoms is in the 1887 catalogue of Peter Van Schaack and Sons, a Chicago wholesale drug supply house. Their catalogue for that year lists condoms under the term capotes; white rubber capotes went for seventy-five cents per dozen while a dozen pink cost only fifty cents. Also listed were rubber caps which were slightly less per dozen.[20]

John Peel, in his study of the English market, found that after 1890, contraceptives were widely available in England, mostly made of rubber, and included rubber letters, skin letters (condoms), American tips, prolapsus check pessaries, Mensinga pessary with spring rim, gem pessary with sponge dome, spring unique pessary with solid or inflated rim, inflated ball pessaries, combined pessary and sheath, and various kinds of syringes, as well as the Rendell soluble pessary and various chemical compounds.[21] The same was probably true in America, although the devices listed in the Sears Roebuck catalogue at the turn of the century which could be used for "contraceptive" purposes were limited to syringes and sponges.[22]

There were many technical problems with contraceptives, one of the main ones being quality control, since there was neither patent nor copyright protection. None of the major rubber manufacturers, at least as indicated by the archives at Akron University, manufactured contraceptives, and the market was left to a number of smaller companies, some of them with a very tenuous financial base. Eventually several companies with adequate quality control such as Youngs Rubber, Julius Schmid, and J. Akwell emerged, but their manufacturing problems were difficult until well into the twentieth century. When the major manufacturing companies did enter the contraceptive field, they often set up separate companies such as Ortho Pharmaceutical Corporation (a division of Johnson and Johnson) or took over one of the older, established companies. In spite of these difficulties, contraceptives were fairly ubiquitous by 1900. They were being used by increasing numbers of women, and their usage helped women to mount a challenge to traditional male domination.

III

Equally important, and also closely associated with rubber technology, was the development of the infant feeding nipple. Without entering into the current controversy over the merits or demerits of breast feeding, it seems obvious that in the past the whole process has been regarded as somewhat confining and restrictive. Some indication of this is the historical ubiquity of wet nurses and the search for other substitutes.

As early as the sixteenth century, there were some areas of Europe where breast feeding was actively discouraged. Infants in such areas were fed pap meal instead of the breast, but the resulting mortality rates were extremely high. [23] The bubbly bottles and other developments were no less dangerous to the young infants. Wet nurses were not that much better. Some evidence for this comes from a 1971 study of twins in Senegal, where 47 per cent of the twins (breast-fed) did not survive beyond their first birthday, compared to 27 per cent of the non-twin population. [24] Since the wet nurse usually had an infant of her own to feed, the Senegal figures for twins became relevant. In a significant number of cases the wet-nursed infant must have died. Shorter collected data on wet nursing in the eighteenth century, and infant mortality figures sometimes approached 90 per cent, depending on the economic class of the parents.[25] Informal studies of wet nurses in Egypt during the 1966-1967 period would tend to replicate the higher range of mortality found in the eighteenth century.[26] In spite of this, large numbers of women were reluctant to nurse their own infants, and wet nurses were standard in most of the upper-class families of Europe and America. Real freedom from the biological fact of nursing had to await alternatives which would make males as capable of infant care as females. This occurred in the last part of the nineteenth century.

Jonathan Gathorne-Hardy found that advertisements for wet nurses appeared regularly in The Times of London in the first half of the nineteenth century, but towards the end of the century there was a rapid decline. From about 1885 onwards the advertisements disappear and wet nurses are neither asked for nor offer themselves. Obviously, wet nurses continued to be used, but they were less likely to be used by the classes reading The Times.[27] Later, when mother's milk was particularly necessary, there was a tendency to turn to special milk collection depots where mothers sold their milk, to be given to a baby through a bottle rather than from the breast.[28]

The reason for the disappearance of the wet nurse is two-fold: the development of the infant feeding bottle, made possible by the rubber nipple, and the pasteurization of milk. Patents for rubber nipples are not hard to find, although the subject has generally been ignored in histories of the rubber industry.[29] Rubber nipples appear almost from the time rubber was vulcanized. The first American patent was granted to Elijah Pratt on August 4, 1845 (U.S. Patent No. 4,131), and others soon followed, including one by the English firm of Maw and Sons. After 1868 there was a literal explosion of patent applications in this area. For example, on June 9, 1868, a patent for a nursing nipple was granted to Francis H. Holt (78,741); in December, 1869, to H. D. Lockwood (No. 97,659); in August, 1871, to McIntosch (No. 118,378); in September, 1872, to George Stevenson, who made improvements to the nipple by dividing it into chambers (No. 131,130). The first seamless nipple was patented by C. B. Dickinson of Brooklyn, New York, in November, 1874 (No. 156,549). It was designed to fit over the mouth of the bottle, and its inventor claimed that removal of the cross strips made cleaning easier. There were patents for nipples on long tubes, a pear-shaped bottle with an imitation rubber

breast which could be worn by the female, and a French-
made bottle with special tubes and ivory pins to regulate
the flow.

One indication of the changing pattern of infant feeding
is the growing concern to deliver better milk supplies. Part
of this concern might well have been due to the realization
that diseases could be transmitted in milk, but this concern
was intensified by the need to get quality milk supplies for
infants and young children, many of whom previously would
have been nursed. One result was the growth of pasteurization and control of milk and, in fact, the emergence of
formalized public health procedures.[30] As a consequence,
infant care became less confining for all women, particularly
when the developments in infant feeding are correlated with
other technological innovations such as washable cotton diapers.

IV

The last technological innovation with which this study
is concerned has to do with menstruation, a subject which
the author has investigated at some length in earlier studies.[31]
The ultimate technological innovation in this area of female
biology was to remove menstruation from the forbidden and
hidden, to reeducate women that it was not debilitating or
dangerous to their intellectual well-being. In earlier studies
by this author, it has been suggested that at least one of the
reasons women wore so many clothes was to cover up any
signs of this feminine "weakness," although there were also
other "physiological reasons" for wearing so many petticoats
and not pants, including the problems associated with urina-
tion and defecation. One of the major problems that women
in the last part of the nineteenth century had with menstruation was dealing with the discharge, and there were a number
of belts and other contrivances on the market designed to
assist in solving this problem. Some indication of the recognition of concern in this area is the number of patents
granted. Between 1854 and 1914 there were at least 20
different patents by both men and women for napkins, catamenial sacks, sanitary supports, menstrual receivers,
monthly protectors, menstrual receptacles, sanitary napkins,
and catamenial supporters.[32] None basically solved the
problems for women, although those that were manufactured
often found difficulties in reaching their public. Johnson
and Johnson in 1896, for example, developed a disposable
sanitary napkin but later withdrew it when it failed to sell
in any quantities.

A major change occurred in 1920 when the Kimberly
Clark Company found itself left with a surplus of cellulose
wadding material made from wood fibers which had been
used for bandaging materials in World War I. The wood
fibers had the advantage of being much more absorbent than
any other material then on the market, and when the company found that army nurses had been using the material for
menstrual pads and were highly impressed by it, they began
manufacturing the first disposable sanitary "napkins." Here
marketing was a problem again, since it was still a forbidden subject, but gradually, even without advertising, the
product caught on despite its cost of ten cents a pad. [33] At
these prices the benefits initially came primarily to the
middle- and upper-class women, but its long-run effect was
revolutionary. When the sanitary pad is coupled with the
advent of the large-scale development of the indoor toilet,
also a product of the twenties, the necessity of wearing large
quantities of clothing disappeared. Dresses could be made
shorter, and even pants could be worn without too great an
inconvenience.

V

In conclusion, the technological innovations of the
last part of the nineteenth and first part of the twentieth
century which tended to overcome some of the physiological
disadvantages of women removed the biological justification
for the restrictive role into which females had been pushed.
Although women's opportunity to participate in the economy
initially may have shrunk with industrialization, since the
reorganization of agriculture tended to displace women, and
artisans' wives who had participated with their husbands
were increasingly less able to do so, [34] the long-term effect was to give greater opportunities to women. These
effects, however, were not felt immediately. Domestic
service offered the major opportunity for female employment all through the last part of the nineteenth century,
and this only began to decline after 1910, coinciding with
the greater biological emancipation of women. Although the
effects of technological innovation in the household undoubtedly
put a greater stress on the middle- and upper-class housewife, perhaps ultimately leading to a loss of status, as Ruth
Schwartz Cowan argued, there was also a growing consciousness that the confinement to the home was not due to any
basically physiological or biological reasons, but to social
and cultural ones. The ultimate result was perhaps to give

impetus to a role challenge since, except for actual child-bearing, none of the other tasks associated with infant care and family rearing roles were strictly biological, and even child-bearing could be planned.

This result was not felt immediately since the socialization process of both men and women was deeply involved, but this cultural lag increasingly led to stresses in the female role models. Since the males were in the dominant position in society and had taken many of the major role models as their own prerogatives, it was necessary to have a frontal challenge before society as a whole would give way, although individually there were undoubtedly many changes in these models. In a sense, the lower economic groups may have benefited more from the first implications of these technological changes than the upper economic classes, if only because the latter had more to lose, and, at least as far as working women were concerned, the lower economic groups never had adopted middle-class concepts entirely. Increasingly, however, there has been a readjustment in the role models of both men and women as the biological justification for such roles has become less and less valid due to technological innovations.

Notes

1. See, for example, W. F. Ogburn and M. F. Nimkoff, Technology and the Changing Family (Boston, Mass.: Houghton Mifflin, 1955); Robert F. Winch, The Modern Family (New York: Holt, 1952); William J. Goode, The Family (Englewood Cliffs, N.J.: Prentice Hall, 1964). Also, Ruth Schwartz Cowan, "The 'Industrial Revolution' in the Home," p. 206 (above).

2. Ruth Schwartz Cowan, "The 'Industrial Revolution' in the Home," p. 206 (above).

3. Ibid., pp. 205-232 (above).

4. See the discussion by Clive Wood and Beryl Suitters, The Fight for Acceptance (Aylesbury, England: Medical and Technical Publishing Company, 1970), p. 10.

5. My wife and I have developed this theme at somewhat

greater length in our study, Bonnie and Vern Bullough, Prostitution: A Social History (New York:
Crown, 1977).

6. See, for example, John E. Knodel, The Decline of
Fertility in Germany, 1871-1939 (Princeton, N.J.:
Princeton University Press, 1974).

7. See James A. Banks, Prosperity and Parenthood: A
Study of Family Planning Among Victorian Middle
Classes (London: Routledge and Paul, 1954).

8. The early history of contraception can be found in
Norman Himes, Medical History of Contraception
(reprinted New York: Schocken Books, 1970).

9. F. A. Wilde, Das Weibliche Gebär-Unvermögen (Berlin:
Nicolai'sche Buchhandlung, 1838).

10. Edward B. Foote, Medical Common Sense (New York:
Published by the author, 1864), p. 380. The book
was copyrighted in 1862 and this date is used in
the paper.

11. W. P. J. Mensinga, Über facultative Sterilität. I.
Beleuchet vom prophlactischen und hygenischen
Standpunkte für Practische aerzte. II. Das Pes-
sarium occlusium and dessen Applikation (2d ed.,
Neuweid und Leipzig; L. Heuser, 1885).

12. John Peel and Malcolm Potts, Textbook of Contraceptive Practice (Cambridge: University Press, 1969),
p. 4.

13. James T. Cooper, Technique of Contraception (New
York: Day-Nichols, 1928), pp. 135-36.

14. See U.S. Patent Office, Specifications of Patents,
April 2, 1878, p. 179, Patent No. 202,037.

15. See the Patent No. 208,883, granted to Trevanion N.
Berlin on October 15, 1878, and Patent No. 141,069,
granted to Otho M. Muncaster on July 22, 1873.

16. This was the claim of Artur Streich, "Zur Geschichte
des Condoms," Archiv für Geschichte der Medizin,
XXII (1929), 208-213.

17. Foote, op. cit., pp. 278-79; John Cowan, The Science of a New Life (New York: Cowan & Company, 1869), p. 110; and J. K. Proksch, Der Vorbaung der venerischen Krankheiten (Vienna, 1872), pp. 50-51.

18. For a brief account of this, see Vern and Bonnie Bullough, "How Rough Red Hands Led to Rubber Gloves," American Journal of Nursing, 70 (April 1970), p. 777. Rubber gloves developed in the surgery of William Stewart Halsted of Johns Hopkins and he recounted the story himself in W. S. Halsted, "Ligature and Suture Material," Journal of the American Medical Association, 60 (May) 13, 1913), p. 1123.

19. In addition to references cited earlier, see John Peel, "The Manufacture and Retailing of Contraceptives in England," Population Studies, 17 (1963-64), 113-125, and The Modern Condom: A Quality Product for Effective Contraception, issued by the Department of Medical and Public Affairs, George Washington University Medical Center, Series H, Number 2, May, 1974.

20. See page 439 of the company's catalogue for that year. A copy can be found in the archives of the American Institute of the History of Pharmacy at the University of Wisconsin. I am indebted to the director, John Parascandola, for furnishing me the information. Generally, as is the nature of such topics, there was little advertising until well into the twentieth century. The Chemist and Druggist of Dec. 1, 1934, for example, included an advertisement for contraceptives from Burge, Warren and Ridgley, which stated that the company had been in the business for over 30 years (p. 24). There was a full-page advertisement for Rendell's soluble pessary in June 27, 1936 (p. 89), in the same journal.

21. Peel, "Manufacture and Retailing," p. 116.

22. Sears-Roebuck Catalogue, 1902, 455-456.

23. John Knodel and Etienne Van de Walle, "Breast Feeding, Fertility and Infant Mortality: An Analysis of Some Early German Data," Population Studies, XXI (1967), 109-131.

24. P. Cantrelle and H. Leridon, "Breast Feeding, Mortality in Childhood and Fertility in a Rural Zone of Senegal," Population Studies, 25 (1971), 505-532.

25. Edward Shorter, The Making of the Modern Family (New York: Basic Books, 1975), p. 181.

26. From personal experiences by Bonnie Bullough, who taught for a time in a Cairo maternity hospital. See Bonnie Bullough, "Malnutrition among Egyptian Infants: A Brief Research Report," Nursing Research 18 (1969), pp. 172-173.

27. See Jonathan Gathorne-Hardy, The Unnatural History of the Nanny (New York: Dial Press, 1973), pp. 36-42.

28. There are still a few of these around. Bonnie Bullough worked in such a station in 1953 in Chicago.

29. See, for example, P. Schidrowitz and T. R. Dawson, editors, History of the Rubber Industry (Cambridge: W. Heffer and Sons, 1952). This book mentions nothing about rubber nipples or contraceptive materials, although it does include an article by G. L. M. Mammon on "Rubber in Medicine and Surgery."

30. For some discussion of this, see George Rosen, A History of Public Health (New York: M. D. Publications, 1958), pp. 344 ff.

31. See Vern Bullough and Martha Voght, "Women, Menstruation, and Nineteenth-Century Medicine," Bulletin of the History of Medicine, XLVII (1973), 66-82. This has been reprinted in Bullough, Sex, Society, and History (New York: Science History Publications, 1976). See also Vern and Bonnie Bullough, Sin and Sickness: Sex in Western Culture (New York: New American Library, 1977).

32. For some patents see No. 11,574 (22 August 1854), catamenial supporter; 57,664 (4 Sept. 1866), catamenial sack; No. 75,036 (8 March 1868), catamenial sack; 75,434, (10 March 1868), catamenial sack; 182,024 (12 Sept. 1876), catamenial sack; 84,874 (15 December 1868) menstrual receiver; 174,540 (7 March 1876), catamenial sack; 235,884 (28 December 1880), catamenial sack; 276,770 (1 May 1883),

monthly protector; 282,201 (21 July 1883), catamenial sack; 296,104 (1 April 1884), catamenial
sack; 297,724 (22 April 1884), catamenial sack;
300,770 (25 June 1884), menstrual receptacle;
393,882 (4 December 1888) catamenial sack; 964,267
(12 July 1910), sanitary napkin and belt; 1,041,420
(15 October 1912), sanitary support; 1,107,447 (18
August 1914), sanitary napkin; 1,158,182 (26 Oct.
1914) napkin.

33. Information received from Kimberly Clark Company.

34. See Ivy Pinchbeck, Women Workers and the Industrial
Revolution, 1750-1850 (London: G. Routledge &
Sons, 1930); Neil Smelser, Social Change in the
Industrial Revolution (Chicago: University of Chi-
cago, 1959); B. L. Hutchins, Women in Modern
Industry (London: G. Bell & Sons, 1915); Edward
Cabury, M. Cecile Matheson, and George Shann,
Women's Work and Wages: A Phase of Life in an
Industrial City (Chicago: 1907); Theresa McBride,
Rural Tradition and the Process of Modernization:
Domestic Servants in Nineteenth Century France,
unpublished Ph.D. dissertation, Rutgers University,
1973.

2. TOYS, TECHNOLOGY AND SEX ROLES
IN AMERICA, 1920-1940

by Carroll W. Pursell, Jr.

I

The decades between the two world wars were characterized, in America, by the marketing of a large number of
consumer durable goods which, in conjunction with contemporary innovations in credit and marketing techniques, transformed the way most people lived. From the automobile,
through the radio, to the electric iron, mass production was
matched by mass consumption of new technical marvels.

With surprising rapidity, the new adult technology
was scaled down for children. Toy vehicles, tools, appliances and construction sets quickly introduced children to
the marvels of owning and using modern technology. Not
surprisingly, the anti-feminist ideology of the adult world--
which dictated which devices were to be used by which sex--
was equally pervasive in the child's world of toys.

Toys depicting modern science and technology have
been ubiquitous in the twentieth century. In describing "A
Child's Christmas in Wales," Dylan Thomas first listed the
useful presents he had received, then described the useless
ones: "Bags of moist and many-colored jelly babies and a
folded flag and a false nose and a tram-conductor's cap and
a machine that punched tickets and rang a bell; never a
catapult; once, by mistake that no one could explain, a little
hatchet.... And Easy Hobbi-Games for Little Engineers,
complete with instructions. Oh, easy for Leonardo!"[1]

Compare this poetic remembrance with a more recent item, written by a woman for the Liberation News
Service:[2]

Every year I remember with a pang the terrible Christmases when I didn't get my chemistry set. My Catholic grammar school never got into any science heavier than the efficiency of a six-day Creation, but my big brother in high school spoke mysteriously of labs and experiments and I was obsessed with a passion to learn.

For two years in a row I desperately wanted that chemistry set and for months before each Christmas I dropped huge hints to my parents and every aunt and uncle in the area. Both years I was bitterly disappointed.

Sometime during the next year I discovered that girls aren't any good at math or science anyway, forgot about the chemistry set and begged for a Harriet Hubbard Make-Up Doll. Of course I got it, and lavished love and lipstick on that doll with all my childish energy. To this day I don't know what makes litmus paper turn colors, but I can apply eyeliner with the best of them.

Even a casual glance at the toys available for children in recent years reveals two striking facts: first, that children are being encouraged to follow the scientific and technological fads of their elders; and second, that these are often advertised as being more appropriate for one sex than for the other. The Gilbert Co., for example had an Ecology kit on the market in 1973, just as it had a wireless telegraph outfit in 1920. At the same time Tonka was marketing a kind of anti-ecology toy: a model snowmobile. Anyone driving the freeways of America soon becomes aware of an explosion of the population of motor homes, and Tonka accepted this challenge by offering the Mighty-Tonka Winnebago. Advertisements showed a little boy apparently driving the vehicle to the campsite, while a little girl took care of the family once they had arrived. One Tonka innovation seemed strangely anachronistic. During the same year (1973) that Charlie Chaplin's classic film "Modern Times" was re-released, and Newsweek reported that "Boredom on the Job" drove 500 of the 4000 workers at one automobile assembly plant to heroin addiction, Tonka unveiled its Assembly Line Kit. The company boasted that each of the cars "can be assembled, played with, taken apart and reassembled. Again and again."[3]

Of course, the traditional technologies continue to be available. The Erector set remains very much the same, and for girls the same company offers Raggedy Ann ironing toys and a play kitchen.   Each of these Queen Size items is so sturdily constructed, we are told, that it "makes a little girl want another Queen Size appliance."[4]

These examples point up several important facts about toys.   First, in all ages and areas, toys have been used not only to amuse and entertain, but also "as socializing mechanisms, as educational devices, and as scaled down versions of the realities of the larger adult-dominated social world."[5] Thus, for example, all Eskimo children play with dolls up to a certain age, then boys throw small harpoons at walruses, seals, and whales carved of bone. [6]

In the United States, studies have indicated that "by the age of three or four, boys and girls show decided preferences for appropriately sex-typed activities, toys, and objects."   It has been pointed out further that among those cultural artifacts of our society which help to form and strengthen patterns of children's play--the media, formal education, direct parental instruction--"none is so constant and concrete in its impact upon children's play as children's toys."[7]

A recent investigation of "Children's Toys and Socialization to Sex Roles" has discovered a large area of agreement between the assumptions of toy manufacturers (the producers and advertisers), parents (the purchasers), and children (the ultimate consumers), concerning male and female sex roles and the toys appropriate to each.[8]   The study concludes that "boy's toys are viewed as the most active, the most social, and relatively high in terms of complexity. Girls' toys, to the contrary, were not seen as 'the most' anything.   They were the least complex (most simple), the least active (most passive), and were virtually tied with the boys' toys in their relatively low ratings on creativity and education."   Some of the more striking data show, for example, that three out of four chemistry sets pictured only boys on the boxes, and the remainder pictured both boys and girls--none showed girls only.   On craft sets, there was a correlation between the sex of children on the boxes and the "high" or "low" technology involved--bead craft showed girls, electronics showed boys.   Boys received more gifts in absolute numbers, their gifts cost more, and were more widely varied, according to this study.

II

A study of American toys marketed during the decades
of the 1920s and '30s leads to the conclusion that this use of
toys to socialize children into what were considered appropriate sex roles was equally prevalent then. This is especially obvious when one looks at those toys which were
thought to embody the principles of modern science and technology.

During these two decades contemporary observers
commented on three basic changes which influenced the world
of toys. First was a dramatic change in technology itself,
especially in the familiar terms of consumer goods. The
airplane, automobile, gasoline tractor, radio, and a host of
electrical appliances became common adjuncts of modern
life. Second, it was widely asserted that formal education
was becoming not only more widespread but more practical
as well. Learning by doing and the democracy of experience
were hailed as ushering in a new generation of bright, pragmatic, flexible, and innovative Americans.[9] And third,
there was a burst of growth within the toy industry itself.
Before the Great War, hand-made German toys had been
prominent on the American market, and toy sales had been
largely seasonal, concentrating on the Christmas trade. By
the end of the twenties, a highly mechanized, aggressively
merchandised American toy industry had grown up to challenge
successfully the imported German product, and the selling
season had been somewhat smoothed out.

In terms of the new technology, girls' toys are most
easily described because they were simpler, fewer in number,
and concentrated in the areas of cooking and cleaning. Such
toys were, of course, not new. A toy kitchen, now in the
Metropolitan Museum of Art, was probably made in New York
in the late eighteenth century, and the Museum of the City of
New York contains a toy stove, utensils, and doll which in
1884 reinforced racially as well as sexually stereotyped roles.

During the 1920s, however, many household tasks
were electrified: between 1924 and 1930 the number of users
of electrical household appliances more than doubled.[10]
Mother's new appliances were quickly scaled down for daughter, so that she too could get used to the joys of living
electrically. An article in the December, 1928 issue of
American Home featured "Grown-up accessories for small
people, " and referred to little girls as either "young house-

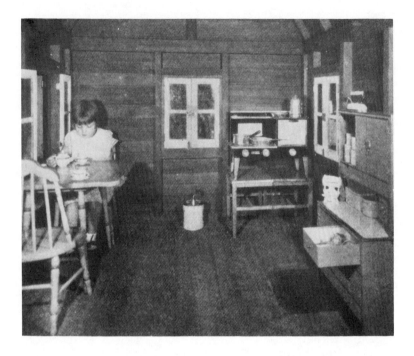

"In this small work room there is an electric stove and iron, kitchen cabinet, drop leaf table and chairs, ice cream freezer and kitchen utensils all of workable, useable size." Christine Frederick, "Grown-up accessories for small people," American Home, 1 (Dec., 1928), 227.

wives" or "small housekeepers."[11] According to the author, "Mothers will welcome a new and interesting development in toys to gladden the heart of the little girl. Boys always have been liberally supplied with outfits and playthings which moved and worked, and which they could use constructively. Such practical gifts have brought boys much fun, because in playing with them they could imitate the many admired activities of 'Father' and other grown-ups. But until recently the small girl has been forced to remain satisfied with tiny dishes, pots, and pans, and with toys of Lilliputian size which she could only pretend were 'just like Mother's.'"

The "young housewife," sitting in her rather isolated (but color-coordinated) kitchen, had a "small scale range

[which] is as perfect in detail and operation as that in her mother's own kitchen. It may be plugged safely into any outlet, and is guaranteed to bake, brew, and brown every-thing from a cake for her father's birthday to a fairy feast for the doll's party. " The various tools and appliances, we are told, "bring happiness because they enable the small girl to exactly counterpart her mother's industries." Such toys, in the words of the article, "satisfy the little girl's natural love of home activities." Since "all little girls love to arrange and rearrange furniture and room furnishings," these too were available, especially "the popular 'dresser' which every true woman adores, no matter her age be six or sixty." F.A.O. Schwarz offered an electric range by 1929, along with the more traditional sewing machine.
Even though the range of toys designed for them was small, girls did on occasion appear in advertisements of boys' toys. But in such cases the girl was seldom active. In one typi-cal advertisement, a girl, clutching her appropriate dolly, is admiring the skill and boldness of her brother, who is playing with a Stanlo set, one of the many construction toys built for boys beginning in this period. 12

Science and technology toys for boys were much more varied, and can be divided into four rough categories: tools, vehicles, construction sets, and science outfits.

Carpentry tools remained the standard item in that field, representing a craft still resistant to change. In 1920 the A. C. Gilbert Co. offered thirteen such sets, ranging from the No. 701 Gilbert Carpenter's Outfit for Boys, at $3.50 to the No. 760 Special Tool Chest. This last item, "built especially for the Government to be used in France by General Pershing's Air Force mechanics during the great war," sold for $50.13 "Extremely well built," according to the catalog, "it makes a chest that any boy can well be proud of."

Closely tied with the gift of tools was their use, and around Christmas time the idea of making toys was popular. A 1934 article in Parent's Magazine gave advice on "Gifts for Them to Make." The author, incidentally, was credited with having written two books--Handicraft for Girls and The Boy Builder. 14 A movement to allow boys to take home economics at school, whiĺe girls could take shop, was en-demic during these years, and in 1932 the vice principal of a high school in South San Francisco reported on one toy-

making experience there.[15] The shop teacher got his boys
into making toys, but as the vice-principal noted, "eighteen
years of tradition had practically 'sealed the fate' of girls
as far as the industrial arts courses were concerned. But
there was a single girl who had nerve enough to talk the
matter over with the principal.... The fact must not be
overlooked," he added, "that this girl was a 'martyr' to the
cause, for she certainly received--well, as high school jar-
gon has it--she received the 'razz.' Only for a few weeks,
however, for because of her earnestness, her nonchalant
attitude, and the 'walking Mutt and Jeff' which she made,
she soon became the object of envy for many girls." The
next semester fifteen girls enrolled in shop--and were taught
in a segregated class.

The category of vehicles was varied and in some
ways very traditional. Wagons, sleds, scooters and bicycles
were still popular, but small cars such as those made by
the Wolverine Co. represented, in 1924, "the latest in juvenile automobile design." As the firm declared, "we conceived the idea that the most outstanding mode of transportation in a modern child's mind was an automobile."[16] Although such vehicles might appear to be relatively free of
sex-orientation, in fact, wagons and such were usually pictured either with the girl being pushed or pulled about, or
in some cases, subordinated by perspective. The Auto-
Wheel Coaster of 1921 was the subject of a concerted sales
campaign aimed at boys. As the catalog predicted, "if Tom
has an 'Auto-Wheel,' you may be sure that Dick and Harry
want one too!" Auto-Wheel Clubs were said to have enrolled
"25,000 boys," all of whom received copies of the Auto-
Wheel Spokes-man, "a lively little publication full of good
live stuff that every boy likes to read." In addition, reflecting the rise of credit buying among adult consumers,
the clubs "help the boys to buy their 'cars' and often actually
advance club funds for this purpose."[17]

On a smaller scale, Gilbert in 1920 offered a toy
tractor stating that "every wide-awake boy knows what
wonders the Tractor has accomplished, and what a tremendous aid it is in the great farming districts of the West.
You boys want to see how these up-to-the minute machines
work."[18]

Even more dramatic was the coming of aviation.
Model airplanes were available from 1904, only months after
the successful first flight of the Wright Brothers. A magazine

reported in 1913 that "boys are now" making flying models
of planes in such numbers that "a national model aeroplane
club is forming from the many local clubs in existence."
The boys, it was claimed, "while profiting in most ways by
the experience of the real bird men and scientific model
builders, work out their own planes from actual experience
in flying the machines." The article concluded that "the
significance of this is evident, for with the coming of the
new industry--as come it must--there will be a great demand for practical aerial engineers. And there is no surer
foundation for success in this line of work than this early
training [sometimes even in school shops] of the boy in the
fundamentals of aviation."[19] In the years immediately after
World War II, the Comet Industries Corporation created
Captain Comet, Jet Ace, to "introduce you to the fascinating
world of aviation." The "you," of course, referred to boys,
as the back cover made explicit with the slogan "Model
Building Builds Model Boys." In a period which valued "to-
getherness" and when "juvenile delinquents" was a phrase
increasingly heard and feared, the Comet company pointed
out that "model building directs youthful energy into con-
structive channels; in the home, it becomes a unifying influence, as father and son work together in building faithful
replicas of famous airplanes."[20]

Science kits, sets, or outfits were also popular with
boys during the twenties and thirties. The 1920 catalog of
the A. C. Gilbert Co., entitled Gilbert Boy Engineer, listed
numerous kits, such as that for civil engineering. "With an
outfit of this kind," Mr. Gilbert wrote, "you are doing something real--something every boy wants to do." Other kits
covered hydraulic and pneumatic engineering, magnetism,
sound, meteorology, machine design, signals and electricity.
The kit for light experiments perhaps was closer to magic
(the field in which Gilbert got his start) than modern science.
Boys were told: "You simply press a button or turn a
switch and you have light. Do you know why--or where it
comes from? No! Because it's electricity."[21]

The most popular kit was probably the chemistry set.
Such sets had been on the market for many years. A
"Boy's Own Laboratory" was offered to the public in 1882,
contained 54 chemicals, 30 pieces of apparatus "for performing endless experiments in Chemical Magic," was warranted "free from danger," and sold for $6.00.[22] World
War I had been called by some a "chemists' war," and the
1920s saw a tremendous growth of the American chemical

industry. In terms of children, the Porter Chemical Co.
made the message explicit in its 1928 catalog of Chemcraft
sets--today boys play at chemistry, tomorrow men hold
scientific posts with industry. "Today," it explained, "no
matter what profession a man follows, he is greatly handicapped without a knowledge of chemistry. The manufacturer,
the farmer, the tradesman, the professional man, the
scientist, all have constant need of chemical knowledge. In
the home the housewife who knows nothing of the chemistry
of the food which she prepares or of the materials which
she uses daily is seriously handicapped."[23]

It was not all serious learning with no fun, however.
Despite a traditional view of the history of science which
celebrated the rise from "the Alchemist of Old to the Modern
Chemist," chemical magic was emphasized. "Chemistry,"
it was claimed, "is also a spectacular science and many
chemical phenomena are most startling and mystifying to the
layman." Facing what must have been a perrenial parental
worry, the firm admitted that "chemistry is sometimes
looked upon as a dangerous profession, but this is not the
case. Contrary to an old popular idea, a chemical experiment does not necessarily result in an explosion."[24] Once
again, there seems to be no good reason why chemistry
sets should have been limited to boys, but advertisements
typically showed only boys interested in the toy.

Such science toys were claimed to have a profound
impact, at least upon some children. The head of one toy
company was said to believe that "very often the careers of
great men in various scientific fields have found their first
inspiration ... in the playthings which amused and fascinated
them in childhood." George Ellery Hale, the American
astronomer, was singled out as a contemporary scientific
luminary who was benignly influenced by toys.[25] One study,
published in 1922, attempted to discover the influence of
after-school science clubs, as the nearest controllable ap-
proximation to free play with science kits. After writing
approvingly of Newton and Faraday as being among "that
galaxy of men who have brought science to its present state
[who] developed their love, interest, yes, and their fundamental background and experience in play with science toys,"
the author concluded that among other benefits, "extra-cur-
ricular activities in science represent a type of purposeful
activity which encourages originality and inventiveness and
habituates boys to the experimental procedure."[26]

Among all categories of toys, those dealing with construction were said to be the most American. In actual
fact, the English Meccano sets of Frank Hornby apparently
antedated American sets. According to an oft-repeated
story, Hornby observed a crane working outside his train
window while on a Christmas trip, and proceeded to develop
toy components which could be assembled into cranes--or a
variety of other machines. His obituary noted that his imagi-
nation was "kindled to the possibility of mechanical toys
which would appeal to the boy mind in general, " and that
this toy "could be put together by any boy of average intel-
ligence. " The notice also recalled that he began his works
in "one room and he had only one girl to assemble the parts. "[27]

By the time he died in 1936, Hornby was publishing
his Meccano instructions in seventeen languages and as early
as 1915 his annual competition for new models was attracting
10, 000 entries. The winner in that year was a toy loom
sent in from New York City. "Meccano, " said the company,
"does teach boys engineering. All the time they are building
models they are acquiring knowledge which may some day
prove of the greatest practical value to them. "[28]

Several American construction toys, dating from this
same period, have remained popular. The Chicago architect
John Lloyd Wright introduced the Lincoln Logs for building
frontier forts, cabins, and other structures. The Lincoln
Log firm also held competitions for new designs, and Mr.
Wright asserted that "a real American boy with a keen brain
is just about the smartest and most original thing alive.
Precedent and custom mean nothing to him. He is bold and
courageous in the execution of his ideas for this reason. "
It was also his belief that "the future salvation of the American toy industry lies mainly in the manufacture of construction toys, these being the kind the foreign makers know the
least about. "[29] In addition, because the construction kits
contained large numbers of identical parts, they could be
mass-produced by machines, reducing costs and the need for
skilled labor.

Another construction toy made in large quantities by
specialized machines, introduced in 1914, was the Tinker
Toy. The president of Toy Tinkers, Inc., in 1924 claimed
that in their factory, "design, plans and preparations, automatic machinery, expert labor, simplification, standardization,
synchronizing of production, all work together to give us a
manufacturing advantage that cheap labor, cheap materials,

and generations of toy-making experience cannot offset."
Tinker Toys, he boasted, had had twenty-nine imitators,
none of whom had been able to operate at a profit. [30]

The leading American construction toy, perhaps, and
the one most like Meccano, was the Erector Set. The toy
was the invention of the remarkable A. C. Gilbert, who was
once described as blending "the familiar qualities of Frank
Merriwell, Theodore Roosevelt, Peter Pan and Horatio
Alger." Frail as a child, Gilbert followed the familiar
regimen of strenuosity, became a star athlete at Yale and
set a world's record in the pole vault in 1908. The next
year he received a medical degree, although he never prac-
ticed. Instead, he began to manufacture apparatus for magic
shows, a childhood passion. Throughout his life, he appears
to have been obsessed with the virtue of manliness and the
need for competition, qualities often associated with science
and especially engineering. "Everything in life is a game,"

Advertisement by A. C. Gilbert, featuring information on his
radio program "Engineering Thrills," his "Look Em-Over"
book, and his contest with 1,021 prizes. The copy carried
Gilbert's assertion that "I am positive every red-blooded boy
will want to be an Erector Engineer this year." Parents'
Magazine, 8 (Dec., 1933), p. 2.

he once reflected, "but the important thing is to win." Like Hornby, he received his inspiration for the Erector set (1914) while looking out a train window at girders being erected. Years after making patterns for the first Erector set, Gilbert remarked that "I've remained a boy at heart and only introduced items that appealed to me. I figured they would appeal to all boys."[31]

Some flavor of Gilbert's thought can be gotten from the introduction to his 1920 catalog: "I feel," he wrote, "that every boy should be trained for leadership. It is only the bright-eyed, red-blooded boy who has learned things, done things, dared things beyond the reach of most boys who will find the way open to really big achievements.... My toys are toys for the live-wire boy, who likes lots of fun and at the same time wants to do some of the big engineering things--things that are real--things that are genuine."[32]

Always alert to what boys might like, Gilbert by 1920 had war toys available. Of the firm's machine gun, the catalog boisterously proclaimed: "if there ever was a real live-wire toy for the red-blooded boys, this is it. Say, you can have more genuine sport with this machine gun than anything I know of. It's the real thing."[33] His "most significant" war toy was also one of the few he marketed for girls--"a nurse's outfit, with cap, arm band, bandages, adhesive tape, scissors, and a bottle of soda-mint tablets."[34] In 1938 Gilbert bought the line of American Flyer trains, and after World War II he brought out an Atomic Energy Laboratory.

The Erector set remained the firm's most famous product, however. Like Meccano, the Gilbert Co. used a wide variety of devices to reach the boy market. During the Christmas season of 1933, for example, the firm's "big illustrated 'Look-Em-Over' Book" contained an entry blank for the Erector model contest: second prize was a new Chevrolet, first prize a trip to "the Panama Canal, or Boulder Dam or the Empire State Building or any other engineering project in the United States." Boys were also urged to tune in to the Sunday evening radio program, "'Engineering Thrills' True stories about real engineers and their hair-raising adventures in digging the Panama Canal, building bridges and sky-scrapers."[35]

From model airplanes to chemistry outfits, from Tinker Toys to Erector sets, it was confidently asserted during the interwar years that such toys helped children,

especially boys, become prepared for life in the world of
modern science and technology. "Our children," wrote the
president of American Flyer in 1921, "are growing up to
manhood and womanhood to face an intense industrial era--
a machine governed world--which will call to its aid the
highest efficiency of science to achieve mass production at
minimum costs." American toys, he concluded, instruct
the child "in the fundamentals of the great mechanical forces
with which he must cope in his adult days."[36] In his autobiography, A. C. Gilbert cited a Yale professor who claimed
that "one of the biggest factors in the growth of the chemical industry in the United States had been Gilbert Chemistry
sets," and cited "the hundreds of letters" he had received
"from engineers" who told him that "their first interest in
their profession started with an Erector set."[37]

III

This education for the modern world of science and
technology was, of course, different for the two sexes.
Household toys, often replicas of the latest electric appliances, were directed at girls. Science and construction
toys, tools and vehicles were aimed at boys. The distinction was obvious and almost without exception. The influence of this ubiquitous training for both boys and girls
to "know their place" would be difficult to measure, but
must have been enormous. The constant urging of boys to
investigate, experiment and innovate was designed to bring
each boy to his highest level of capability. Girls, on the
other hand, were not so urged, but given toys (such as dolls)
which encouraged nurturing, and appliances which prepared
them for a lifetime of cooking, washing and cleaning. As
one recent critic has noticed, while "it is true that men's
options are also limited by our society's sex-role ideology,
it is still the women in our society whose identities are
rendered irrelevant by America's socialization practices."
Perhaps all of this is one reason that a 1965 study discovered
that by the ninth grade, 25 per cent of boys, but only three
per cent of girls, were considering careers in science and
engineering.[38]

Notes

1. Dylan Thomas, A Child's Christmas in Wales (N.Y.,
   1954), pp. 16-17.

2. Karen Kearns, Liberation News Service, December 17, 1969.

3. Tonka Corp., Tonka '73 (n.d.), p. 34.

4. Hubley Division, Gabriel Industries, Inc., Hubley Toys 1973 (n.d.), p. 42.

5. Donald W. Ball, "Toward a Sociology of Toys: Inanimate Objects, Socialization, and the Demography of the Doll World," Sociological Quar., 8 (1967), p. 447.

6. Lola M. Cremeans, "Eskimo Toys," Jour. of Home Economics, 23 (April, 1931), 355-358. For American Indians, see Robert H. Lowie, Indians of the Plains (N.Y., 1963), p. 131.

7. Louis Wolf Goodman and Janet Lever, "Children's Toys and Socialization to Sex Roles," Aug., 1972, unpublished, quoted by permission.

8. Ibid.

9. See, for example, W. Ogden Coleman, "Toys, Moulders of Industrialists," American Industries, 22 (Aug., 1921), 13.

10. Harry Jerome, Mechanization in Industry (N.Y., 1934), p. 432.

11. Christine Frederick, "Grown-up Accessories for Small People," American Home, 1 (Dec., 1928), 227, 276, 278.

12. Ad for Stanlo in Parents' Magazine, 8 (Dec., 1933), 65.

13. A. C. Gilbert Co., Boy Engineering (New Haven, 1920), p. 105.

14. Edwin T. Hamilton, "Gifts for Them to Make," Parents' Mag., 9 (Dec., 1934), 81.

15. On shop and home economics, see Jour. of Home Economics, 8 (April, 1916), 214 and ibid., 22 (Aug., 1930), 659-660. George Carl Weller, "Toymaking for Girls," Education, 53 (Oct., 1932), 112-113.

16. Dail Steel Products Co., Wolverine Juvenile Vehicles (Lansing, 1924), p. 15.

17. Auto-Wheel Coaster Co., The New Models (North Tonawanda, 1921), p. 15.

18. Gilbert Co., Boy Engineering, p. 121.

19. Edward I. Pratt, "Boys as Aeroplane Modelers," Illustrated World, 20 (Nov., 1913), 423-425.

20. Comet Industries Corp., Meet Captain Comet "Jet Ace" (n.p., n.d.), back cover.

21. A. C. Gilbert Co., Boy Engineering, p. 76.

22. Frederick Lowey & Co., Lowey's First Steps in Chemistry (N.Y., 1882).

23. Porter Chemical Co., Chemcraft No. 10 Experiment Book (Hagerstown, 1928), p. 11.

24. Ibid.

25. Alfred Albelli, "Toys Make the Man," Popular Mechanics, 52 (Dec., 1929), 962-963.

26. Morris Meister, "The Educational Value of Scientific Toys," School Science and Mathematics, 22 (Dec., 1922), 801-813.

27. Obituary notices of Hornby in New York Times, Sept. 22, 1936, and The Times (London), Sept. 22, 1936.

28. Meccano Co., Meccano Prize Models: A Selection of the Models Which Were Awarded Prizes in the Meccano Competition 1914-15 (N.Y., 1915), p. 13.

29. G. A. Nichols, "How Advertising Opened All-Year Market for Toys," Printer's Ink, 123 (April 26, 1923), 124.

30. The President, The Toy Tinkers, Inc., "Manufacturing Policies that Have Offset Europe's Cheap Labor," Factory, 33 (Dec., 1924), 779-781, 846.

31. Obituary notice of Gilbert in New York Times, Jan. 25,

Carroll W. Pursell, Jr. / 267

1961. For his autobiography, see A. C. Gilbert and Marshall McClintock, The Man Who Lives in Paradise (N. Y., 1954).

32. Gilbert Co., Boy Engineering, p. 3.

33. Ibid., p. 120.

34. Gilbert and McClintock, Man Who Lives in Paradise, p. 159.

35. Parents' Magazine, 8 (Dec., 1933), 2.

36. Coleman, 13.

37. Gilbert and McClintock, Man Who Lives in Paradise, pp. 158-9, 144.

38. Judith M. Bardwick, Psychology of Women: A Study of Bio-Cultural Conflicts (N.Y., 1971), p. 117; Sandra L. and Daryl J. Bem, "Case Study of a Nonconscious Ideology: Training the Woman to Know Her Place," in Daryl J. Bem, ed., Beliefs, Attitudes, and Human Affairs (Belmont, 1970), p. 96. Bardwick suggests that "the toys and tasks of boys may foster objective, independent criteria of success; if an erector-set model works it is largely irrelevant what people say although it is nice to hear praise. Whether or not the decoration of a doll house is pretty really does depend upon the judgement of others," p. 117. Such a claim for Erector sets certainly conforms to the tenets of engineering ideology (engineers deal only with objective truth), but the difference is actually one of degree, not kind.

NAME INDEX

SUBJECT INDEX